8th European Summer School on Experimental Nuclear Astrophysics

Big Bang and Stellar Nucleosynthesis, Plasmas in Stars and Laboratories, Direct and Indirect Measurements, Detectors and Facilities for Nuclear Astrophysics, Experiments with RIB

Journal of Physics: Conference Series Volume 703

Santa Tecla, Italy
13 – 20 September 2015

Editors:

Spitaleri Claudio
Lamia Livio
Pizzone Rosario Gianluca

ISBN: 978-1-5108-2284-9
ISSN: 1742-6588

Printed from e-media with permission by:

Curran Associates, Inc.
57 Morehouse Lane
Red Hook, NY 12571

CURRAN ASSOCIATES INC.
proceedings
.com

Some format issues inherent in the e-media version may also appear in this print version.

Printed by Curran Associates, Inc. (2016)

For permission requests, please contact the Institute of Physics
at the address below.

Institute of Physics
Dirac House, Temple Back
Bristol BS1 6BE UK

Phone: 44 1 17 929 7481
Fax: 44 1 17 920 0979

techtracking@iop.org

Additional copies of this publication are available from:

Curran Associates, Inc.
57 Morehouse Lane
Red Hook, NY 12571 USA
Phone: 845-758-0400
Fax: 845-758-2633
Email: curran@proceedings.com
Web: www.proceedings.com

Table of contents

Volume 703

8th European Summer School on Experimental Nuclear Astrophysics (Santa Tecla School)
13–20 September 2015, Santa Tecla, Italy

Accepted papers received: 16 March 2016
Published online: 14 April 2016
Preface

Papers

Lectures

Contributions by young researchers

The 8th European Summer School on Experimental Nuclear Astrophysics
September 13-20, 2015
Santa Tecla Palace Hotel, Sicily, Italy
Big Bang and Stellar Nucleosynthesis, Plasmas in Stars and Laboratories, Direct and Indirect Measurements, Detectors and Facilities for Nuclear Astrophysics, Experiments with RIB
Director of the School
C. Spitaleri (Catania)
Scientific Committee
M. Aliotta (Edinburgh)
M. Busso (Perugia)
A. Coc (Orsay)
M. El Eid (Beirut)
T. Kajino (Tokyo)
K.L. Kratz (Mainz)
S. Kubono (Tokyo)
K. Langanke (Darmstadt)
Wei Liping (CIAE)
J. Josè (Barcelona)
T. Motobayashi (Tokyo)
A. Mukhamedzhanov (TAMU)
E. Nappi (Bari)
O. Straniero (Teramo)
G. Rogachev (TAMU)
C. Rolfs (Bochum)
L. Trache (Bucharest)
R. Tribble (BNL & TAMU)
M. Wiescher (South Bend)
Local Committee
R.G. Pizzone (chair)
G. Agnello
S. Palmerini
V. Potenza
G.G. Rapisarda
S.M.R. Puglia
M.L. Sergi
R. Spartà
Scientific Secretary:
L. Lamia
Contact:
Email: astro2015@lns.infn.it
Website: https://agenda.infn.it/conferenceDisplay.py?confId=8678
INFN
SICILIAE STVDIVM GENERALE 1434

Journal of Physics: Conference Series **703** (2016) 011002 doi:10.1088/1742-6596/703/1/011002

Welcome address

Prof. Claudio Spitaleri, director of the school

It is my pleasure, as co-director of the school, to welcome students and lecturers of the "Eigth European Summer School on Experimental Nuclear Astrophysics". The school is at its seventh edition and we have students coming from several European countries and from all the world, like Lebanon, India, USA, Russia, Canada, Japan, China, Brazil, Kazakhstan. This is a clear sign that the school is considered as an important meeting for the international Nuclear Astrophysics Community worldwide. We are very proud to offer, for the eigth time, our school program, which aims at widening the students' perspectives and sets, hopefully, a milestone in the education of a new generation of scientists. I remind you that among the goals of the school there are two particular points that deserve attention. The first is allowed by the special venue of the school: the program takes place inside a unique building and this allows all the students to continue their scientific activity and the discussions on the lectures beyond the normal lecture hours. The second peculiar feature is that of favouring cultural exchanges between lectures and students with different traditions, customs and religions. Indeed, Sicily has been historically considered as a meeting place for different cultures, a land that offer chances for peaceful and constructive encounters among different peoples.

I wish to thank all the supporting institutions and sponsors, such as INFN - LNS and the University of Catania. To you all I wish a pleasant and fruitful time in our wonderful city.

Dr. Giacomo Cuttone, Director of INFN-LNS

As a nuclear physicist and as the Director of the LNS it is a pleasure to welcome the students and lecturers of the school in our laboratory. For the eigth time Catania assumes the role of an attraction point in the field of Nuclear Astrophysics, since the whole scientific world is represented here. As a scientist I wish you a fruitful attendance and especially to the youngest ones I wish for them to maintain for all their lifetimes the same enthusiasm they show in this moment. The human understanding of the Universe needs also your efforts as Nuclear Astrophysicists to proceed. LNS has an important and active group of Nuclear Astrophysics research and constitutes a leading facility worldwide in this field. I wish you a fruitful and pleasant week in Santa Tecla and Catania.

Preface: Eighth European Summer School on Experimental Nuclear Astrophysics

In this book a collection of the lecture notes given during the Eighth European Summer School on Experimental Nuclear Astrophysics is given. The school, whose first edition was first held in 2001, took place from 13 to 20 of September 2015 in Santa Tecla, a small village about 15 km north of Catania, characterized by its position on the volcanic shores of the Ionian Sea, surrounded by the spectacular "Timpa" area, a green protected park specific for its mediterranean vegetation. 80 young students and researchers from more than 20 countries attended the lectures and were also encouraged to present their work and results.

The school, has tried once more to present to the young students the global picture of nuclear astrophysics research in the last years. Thus the scientific program of the school covered a wide range of topics dealing with various aspects of nuclear astrophysics, such as stellar evolution and nucleosynthesis, neutrino physics, the Big Bang, direct and indirect methods and radioactive ion beams. Nuclear astrophysics plays a key role in understanding energy production in stars, stellar evolution and the concurrent synthesis of the chemical elements and their isotopes. It is also a fundamental tool to explain the ashes of the early universe, to determine the age of the universe through the study of pristine stellar objects and to predict the evolution of the Sun or Stars. The "bone structure" for the above aspects is based on nuclear reactions, whose rates need to be determined in laboratories. Although impressive progress has been made over the past decades, which was rewarded by Nobel prizes, several open questions are still unsolved, which challenge the basis of the present understanding.

A list of the lecture topics is given below:

- Big Bang Nucleosynthesis
- Stellar evolution and Nucleosynthesis
- radioactive ion beams
- detector and facilities for nuclear astrophysics
- indirect methods in nuclear astrophysics
- plasma physics

An often quoted statement of Carl Sagan, "We are all star stuff," is exhaustively describing one of the great discoveries of the 20th century as well as the main aim of nuclear astrophysics. The present theory of stellar nucleosynthesis (based on the famous paper B^2FH) predicts chemical evolution of the Universe, which is testable by looking at stellar spectral lines, meteorites, pre-solar grains or other means of investigation. Quantum mechanics explains why different atoms emit light at characteristic wavelengths, and so by studying the light emitted from different stars, one may infer the atmospheric composition of individual stars. However, upon undertaking such a task, observations indicate a strong correlation between a star's heavy element content (metallicity) and its age (red shift).

Big bang nucleosynthesis tells us that the early universe consisted of only the light elements, and so one expects the first stars to be composed of hydrogen, helium, and lithium, the three lightest elements. The recent achievements of WMAP and Planck missions, with the precise measurement of many cosmological parameters, have re-ignited the interest on the primordial nucleosynthesis, especially for the still unclear Lithium primordial abundance.

Stellar structure and the H-R diagram indicate that the lifetime of a star depends greatly on its initial mass and chemical composition, so that massive stars are very short-lived, and less massive stars are longer-lived. As a star dies, nuclear astrophysics argues that it will enrich the interstellar medium with "heavy elements" (in this case all elements heavier than lithium, the third element), from which new stars are formed. This account is consistent with the observed correlation between stellar metallicity and red shift.

The theory of stellar nucleosynthesis would not be very convincing if reliable nuclear physics inputs are adopted. By carefully scrutinizing the table of nuclides, nuclear astrophysicists were able to predict the existence of different stellar environments which could produce the observed isotopic abundances, and the nuclear processes which must occur in these stars. This is also valid for elements heavier than iron production, when fusion reactions are of negligible importance and the main nucleosynthetic path goes through p-process, r-process, and s-process.

Many of the nuclides generated in the huge explosion triggered by the last burnings in the life of a massive star are unstable. Then in order to understand what is going on in Supernovae explosions, Gamma-ray bursts, Novae it was pointed out that the physics of radioactive beams should be explored. Many group of physicists have then taken the opportunities offered by the world-wide developments of radioactive ion beams facilities to start to investigate such environments. In this field also the weak interaction processes are important and in some cases, like type II Supernovae, dominant. Additional nucleosynthetic paths, such as the i-processes were also introduced in one specific lecture.

In order to better understand our Universe, a precise knowledge of nuclear inputs for nuclear astrophysics is therefore needed. They influence sensitively the nucleosynthesis of the elements in the early stages of the universe and in all the objects formed thereafter and control the associate energy generation, neutrino luminosity and stellar evolution. Thus a good knowledge of reaction rates is essential to understand this broad picture. Studies at the energies typical of nuclear astrophysics for charged particle induced reactions (few keV - 1 MeV) are severely hampered by the low signal-to-noise ratio, essentially due to the tiny cross sections that have to be measured (as low as picobarn in some cases). The best experimental solutions to this main problem will be discussed in this book and presented in their details. The pilot project, LUNA, i.e. a 50 keV accelerator installed in the Laboratori Nazionali del Gran Sasso has already given several important results in the last decade. Other experimental methods such as the recoil separator, spectrometers will be reviewed together with the wide field of nuclear astrophysics research opened by the development of radioactive ion beams in several laboratories of the world.

Detectors and facilities useful for the next decade of nuclear astrophysics studies were reviewed in the lectures and attention was devoted to future research possibilities for studies. This knowledge and the facilities will be of groundbreaking importance for solving the future quests for this field.

Although all these efforts in some cases (such as the $^{12}C +^{12} C$ interaction and the $^{12}C(\alpha,\gamma)^{16}O$ and many others) the goal is far from being achieved by means of direct measurements. Moreover the presence of the electron screening effect makes very hard and sometimes impossible the measurement of the bare nucleus astrophysical factor, that is the quantity needed for astrophysics. Thus extrapolations (usually supported by R-matrix theory) are adopted. An alternative way to by-pass these problems and measure the bare nucleus astrophysical factor or cross section is given by the indirect methods. Among them the Coulomb Dissociations, the ANC, decay spectroscopy and the Trojan Horse Method will be reviewed.

In recent years, new possibilities of exploring the universe have opened. Not only electromagnetic spectra, luminosities are available for stars, galaxies and compact objects but new windows are opened in particle detection. Cosmic ray detectors will play, in the near future, an important role for a better understanding of what is going on in the universe and therefore they will trigger the opening of a new sector of nuclear astrophysics research. Physics experiments can be performed in laser-induced plasmas available worldwide, opening new possibilities for future research.

Crucial problems related to various issues of nuclear as well as astrophysics are still open, triggering new researches in this field and the enthusiasm to organize events such us our European Summer school on experimental nuclear astrophysics. Once more students, young researchers and lecturers have gathered in Santa Tecla from more than twenty countries of the world. The school, though a European one, keeps playing the role of a meeting point of students of different nationalities from the whole world. These students, we are sure, have left our school not only with an improved knowledge of nuclear astrophysics but also with new friends (as well as future collaborators) all around the world.

We gratefully acknowledge all the participants the lecturers and the members of the Scientific committee.The school could not have taken place without their efforts and help. The Organizing Committee also acknowledges the financial support of the Laboratori Nazionali del Sud - Istituto Nazionale di Fisica Nucleare - Italy under the grant "LNS Astrofisica Nucleare (fondi premiali)" , Dipartimento di Fisica e Astronomia of the University of Catania - Italy. All papers published in this volume of Journal of Physics: Conference Series have been peer reviewed through processes administered by the Editors. Reviews were conducted by expert referees to the professional and scientific standards expected of a proceedings journal published by IOP Publishing.

The school has been honoured by the welcome addresses of Dr. Giacomo Cuttone, Director of Laboratori Nazionali del Sud and by the school director prof. Claudio Spitaleri. We wish to the reader that the reading of this book can be as fruitful as the participation to our school.

Claudio Spitaleri
Dipartimento di Fisica e Astronomia, University of Catania

Livio Lamia
Dipartimento di Fisica e Astronomia, University of Catania

Rosario Gianluca Pizzone
INFN - Laboratori Nazionali del Sud, Catania

Journal of Physics: Conference Series **703** (2016) 011004 doi:10.1088/1742-6596/703/1/011004

Journal of Physics: Conference Series **703** (2016) 011005 doi:10.1088/1742-6596/703/1/011005

Peer review statement

All papers published in this volume of *Journal of Physics: Conference Series* have been peer reviewed through processes administered by the proceedings Editors. Reviews were conducted by expert referees to the professional and scientific standards expected of a proceedings journal published by IOP Publishing.

Journal of Physics: Conference Series **703** (2016) 012001 doi:10.1088/1742-6596/703/1/012001

New insights from cosmic gamma rays

Roland Diehl

Max Planck Institut für extraterrestrische Physik, D-85748 Garching, Germany

E-mail: rod@mpe.mpg.de

Abstract. The measurement of gamma rays from cosmic sources at ~MeV energies is one of the key tools for nuclear astrophysics, in its study of nuclear reactions and their impacts on objects and phenomena throughout the universe. Gamma rays trace nuclear processes most directly, as they originate from nuclear transitions following radioactive decays or high-energy collisions with excitation of nuclei. Additionally, the unique gamma-ray signature from the annihilation of positrons falls into this astronomical window and is discussed here: Cosmic positrons are often produced from β-decays, thus also of nuclear physics origins. The nuclear reactions leading to radioactive isotopes occur inside stars and stellar explosions, which therefore constitute the main objects of such studies. In recent years, both thermonuclear and core-collapse supernova radioactivities have been measured though ^{56}Ni, ^{56}Co, and ^{44}Ti lines, and a beginning has thus been made to complement conventional supernova observations with such measurements of the prime energy sources of supernova light created in their deep interiors. The diffuse radioactive afterglow of massive-star nucleosynthesis in gamma rays is now being exploited towards astrophysical studies on how massive stars feed back their energy and ejecta into interstellar gas, as part of the cosmic cycle of matter through generations of stars enriching the interstellar gas and stars with metals. Large interstellar cavities and superbubbles have been recognised to be the dominating structures where new massive-star ejecta are injected, from ^{26}Al gamma-ray spectroscopy. Also, constraints on the complex interiors of stars derive from the ratio of ^{60}Fe/^{26}Al gamma rays. Finally, the puzzling bulge-dominated intensity distribution of positron annihilation gamma rays is measured in greater detail, but still not understood; a recent microquasar flare provided evidence that such objects may be prime sources for positrons in interstellar space, rather than distributed nucleosynthesis. We also briefly discuss the status and prospects for astronomy with telescopes for the nuclear-radiation energy window.

1. Introduction

The nuclear processes in cosmic objects produce, among others, unstable, radioactive isotopes (from nucleosynthesis), or nuclei excited above their ground states (from energetic collisions). Therefore, a measurement of cosmic gamma rays makes uses of electromagnetic radiation (i.e. *astronomy*) as a remote probe to learn about such nuclear processes in distant sources. This complements studies of presolar grains, which are being found as inclusions in meteorites, and provide a material messenger from cosmic nuclear reactions from sources outside the solar system. Together, these are the key tools for nuclear astrophysics for the study of nuclear reactions and their impacts on objects and phenomena throughout the universe. All other studies rely on less direct consequences of nuclear reactions: Supernova light and spectra, for example, originate from the photosphere of a supernova, which is rather remote from the inner energy source, and complex radiation transport links the observables to their physical origins. Nevertheless, astronomical precision is more developed at wavelengths longer than those of this

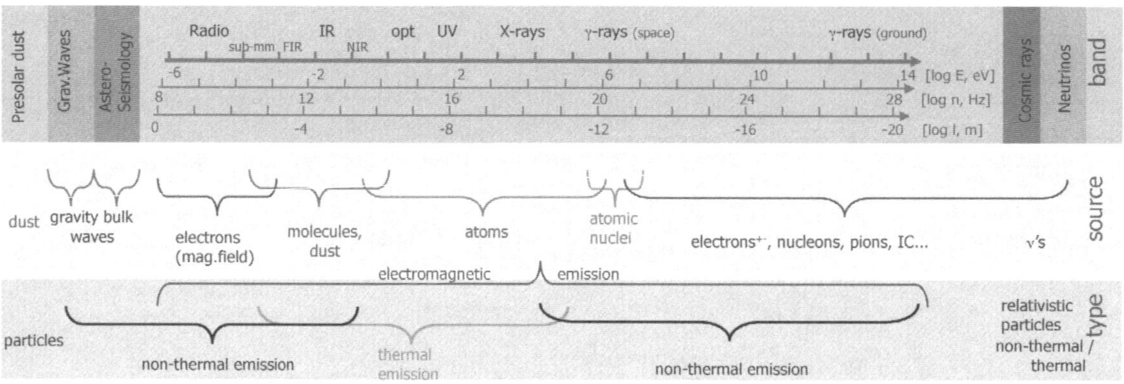

Figure 1. The electromagnetic spectrum covers about 20 decades of photon energy. Other astronomical messengers are listed on the sides. The different physical processes per energy are indicated below. Nuclear processes are reflected by photons in the two decades of 0.1 to 10 MeV only.

gamma-ray range, when astronomical instrumentation is capable to collect and focus radiation that is less penetrating than gamma-rays.

This paper may be seen as a complement to a similar report written ten years ago from a corresponding lecture series at the Santa Tecla school [Diehl, 2007], and to a recent review article [Diehl, 2013a]. Further background material may be found in a recent textbook [Diehl et al., 2011].

2. Nuclear astronomy in the context of astrophysics messengers

Measurements of nuclear radiation and its properties from cosmic sources is difficult. After pioneering detectors on the Apollo spacecraft and in the context of first explorations of the near earth space and solar radiation, it had become clear that radiation at nuclear energies was an observational fact, and that it presented new astronomical constraints [Trombka et al., 1978; Trombka and Fichtel, 1983], such as the *MeV bump* of the diffuse gamma-ray background, line signatures in the high-energy emission of solar flares, and gamma ray bursts. The first Galaxy-scale sources seen in such gamma rays were the lines attributed to positron annihilation reported after a balloon experiment from the central Galaxy in 1964 and the line from ^{26}Al radioactivity gamma-rays also from the central Galaxy region reported from the 1978/79 measurements with the HEAO-C satellite [Mahoney et al., 1982]. But it took many years to further advance the techniques of gamma-ray telescopes, mainly due to the penetrating nature of the gamma rays, which prevent photon collection and concentration by mirrors or lenses, and due to activation by cosmic rays of the instrument materials to generate a large instrumental background [Bertsch et al., 1988]. Therefore, even after two major observatory missions for imaging and spectroscopy [Diehl, 2013a; Gehrels et al., 1993; Schoenfelder et al., 1996; Vedrenne et al., 2003; Winkler et al., 2003], currently achieved sensitivities allow to only see Galactic sources and exceptionally-bright supernovae from up to few Mpc distance, and spatial resolution is of order 2-3 degrees only. The spectral resolution, however, has been advanced to and beyond astrophysical needs and expectations, allowing line shape constraints in detail for the lines and sources of interest [Diehl, 2013b].

The nuclear window of astronomy (see Fig.1) may be defined between the lowest-energy nuclear transitions (of ^{44}Ti at 68 and 78 keV, and ^{60}Fe at 58 keV) and the highest energies of such transitions from cosmic sources of $\sim$15 MeV (more specifically, the ^{16}O transitions at

Journal of Physics: Conference Series **703** (2016) 012001 doi:10.1088/1742-6596/703/1/012001

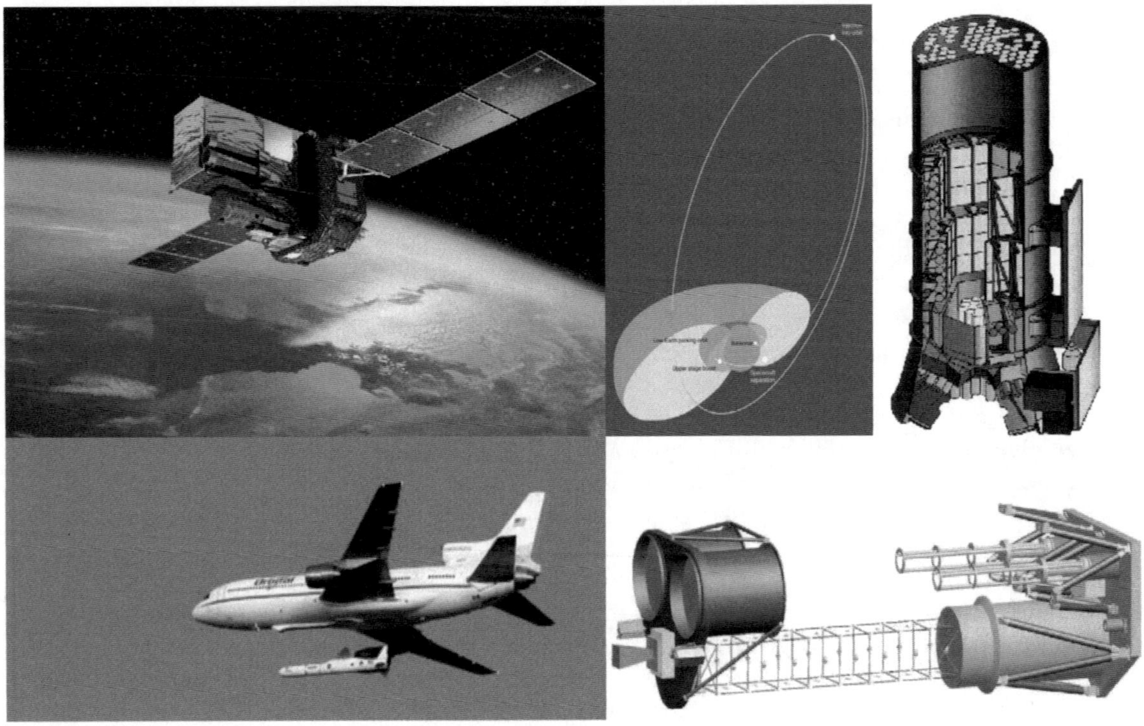

Figure 2. The current instrumentation measuring cosmic photons in the nuclear radiation window. ESA's INTEGRAL mission [Winkler et al., 2003] was launched in 2002 and is currently in its extended mission phase after its 5-year nominal mission *top*; also its excentric orbit is shown *middle* to minimise background variations. The *righthand* figure shows the spectrometer SPI. The NuSTAR X-ray telescope [Harrison et al., 2013] with a focusing mirror on a boom (*bottom*) operates at energies up to 80 keV and was launched from an air plane with a Pegasus rocket in 2012, and also is in its extended phase after its nominal 2-year mission.

6.1 and 7 MeV and the ^{12}C transition at 15.1 MeV). This window also includes the signatures of the annihilation of positrons with its prominent line at 511 keV, which, strictly speaking, is not of *nuclear* origin, but unique in its astrophysical information about high-energy sources and processes. Exploration of this astronomical window is fairly recent, starting ~50 years ago [Lingenfelter and Ramaty, 1978]. INTEGRAL for the first time combines adequate spectral resolution for line identification and line shape measurements with all-sky exposure.

Other information about such high-energy processes of interest are also not easy to obtain: Meteoritic inclusions have been found to contain a wealth of isotopic information [Clayton and Nittler, 2004]. This derives from measurements of material probes in the laboratory with exquisite precision. The results can be related to sources of nucleosynthesis, yet the formation of presolar dust grains and the journey of these through meteorites to our terrestrial laboratory leaves some systematic issues and uncertainties that limit such interpretation. X-ray spectroscopy also addresses high energy physics and abundances of nuclei in hot gas near nucleosynthesis events. But here potential confusion with atomic re-formation processes and the corresponding astrophysical processes, as well as underlying thermal emission, make it difficult to extract nucleosynthesis information. Penetration to dense interiors of cosmic sources of interest is partially possible with gamma rays, more than at X-rays or even lower energies, but still severely affected by radiation re-processing e.g. in stars and supernovae, in particular in comparison with neutrinos. This has been demonstrated in recent years with detailed spectroscopy of solar neutrinos [BOREXINO Collaboration et al., 2008, 2014]; but the Sun may remain the only source for which neutrino spectroscopy at nuclear physics detail is feasible. Neutrino and gravitational-wave signatures are among the hopes of astrophysicists to make use of new messengers in order to learn more about the interiors of stars, stellar collisions, and explosions. But learning about the characteristic signatures in new observational windows, and also the biases and systematic effect, is tedious and takes decades, typically. The astronomical window of nuclear radiation is now explored since about 2–3 decades; in spite of its restrictions, it will remain worth exploring through improved instrumentation.

3. Science quests and lessons for nuclear astronomy

Nuclear-physics processes in cosmic sites are mainly the nuclear reactions in sources of cosmic nucleosynthesis. Inner regions in stars, their core in all cases, and shells for stars beyond the main sequence phase of their initial evolution, but also stellar explosions are prime objects for science investigations using nuclear observations. Furthermore, cosmic sites where high-energy reactions occur or which are cosmic extremes of gravity or electromagnetism have become prominent objects of astrophysical studies also making use of nuclear emission processes.

Interiors of stars are opaque and can be studied only indirectly; but in particular their later evolution beyond core hydrogen burning is plagued with substantial uncertainties [Jones et al., 2015; Pignatari et al., 2013a] from complex hydrostatical equilibrium rearrangements that involve material transport with intrinsic 3-dimensionality such as convection and radiation transport through materials of varying densities and composition. The consequences of this late stellar evolution are significant, because late interior evolution synthesises many nuclei which contribute to cosmic metal enrichment, e.g. from the s-process in giant stars whose products are ejected through winds and major contaminations of dust grains found in stardust inclusions of meteorites [Pignatari et al., 2010, 2013b]. Also, the core sizes of later pre-supernova stages are determined in this late evolution, as are the carbon to oxygen ratio. Not knowing all this interior complexity, models of stellar evolution carry systematic uncertainties which a.o. result in uncertainties of lifetimes till core collapse supernova or strong wind phases.

3.1. Supernovae

The bright display of supernovae, rising within a few days to a maximum luminosity that rivals galaxies, and falling of slowly at a time scale of months, is due to energy input from radioactive decay [Clayton et al., 1969]. The ^{56}Ni produced in the supernova decays through gamma-ray and positron emission, and heats the supernova envelope to provide such a spectacular astronomical phenomenon.

Observing gamma rays from the primary energy source (see Fig. 3) has always been a goal, as well as a dream, for gamma-ray astronomy [Clayton et al., 1969]. The brightest such sources occur quite rare in our nearby universe, and are supernovae of type Ia which produce about 0.5 $M_\odot$ and thus about ten times as much ^{56}Ni as core-collapse supernovae do. In January 2014, luckily one such event occurred at just 3.5 Mpc distance [Fossey et al., 2014], in the nearby star-forming galaxy M82.

INTEGRAL invested almost half a year of observing time to follow the expected brightening of this supernova in ^{56}Co decay gamma-rays, which are expected to leak out of the supernova after several weeks towards a maximum brightness expected about three months after the explosion. From the observations, indeed, Doppler broadened lines at 847 and 1238 keV were measured [Churazov et al., 2014; Diehl et al., 2015] (see Fig. 4). This is a great triumph of supernova science, and confirms our general understanding of a white dwarf being disrupted by Carbon fusion to power the explosion. The brightness of the lines corresponds to about 0.5 $M_\odot$ of ^{56}Ni produced, as expected.

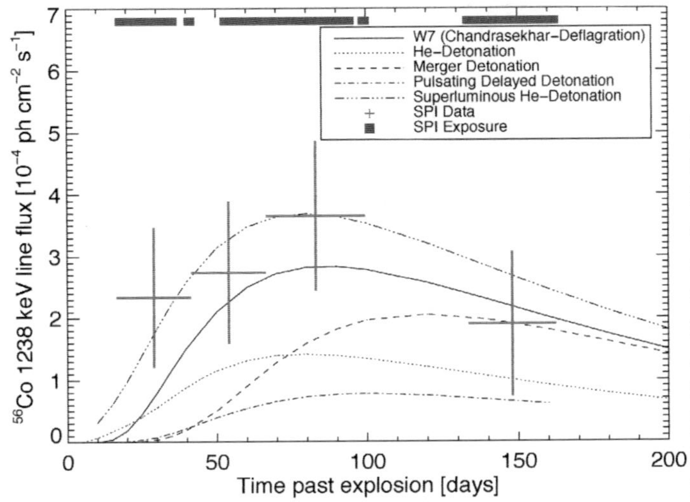

Figure 3. The escape of gamma rays from ^{56}Co decay in a SN Ia is expected to peak only 60–120 days after the explosion, while the optical brightness maximum occurs typically only 20 days past explosion. Here, several model variants are shown (*continuous and dashed lines*)[The and Burrows, 2014], together with data points from SN2014J measured with INTEGRAL for the line at 847 keV [Diehl et al., 2015].

There was a surprise, however: INTEGRAL started to observe the supernova 17 days after the inferred explosion date of Jan 14.75 2014 (UT). This seemed too late to test for presence of ^{56}Ni on the surface, as would be expected by the somewhat exotic *He cap* model variant of SN Ia, considering the decay time of ^{56}Ni of 8.8 days. But INTEGRAL data provided a clear indication of the presence of the characteristic ^{56}Ni decay lines at 158 and 812 keV energy [Diehl et al., 2014]. This was unexpected, and points to a rather significant mass of ^{56}Ni near the surface, as much as 10% of the total. Such constraints are important to find out what the progenitors of supernovae of type Ia are, and how the explosion actually occurs [Hillebrandt et al., 2013]. Therefore, it was hypothesised that the explosion might have been triggered by a surface helium explosion from a helium belt that has built up rather rapidly from matter accreted from the companion star, and thus that the companion also is rather evolved and sheds helium, rather than having a hydrogen envelope[Diehl et al., 2014]. This makes SN2014J appear as an intermediate between a standard white-dwarf / main-sequence binary and a double degenerate system of two white dwarfs. The amount of ^{56}Ni and other constraints, however, suggest that the mass of the disrupted star was rather high and near the Chandrasekhar limit. The narrowness of the observed gamma-ray lines then suggests that our aspect of the SN2014J progenitor binary system is rather face-on than edge-on.

In many ways this supernova appears to be rather *standard*, but nevertheless shows signs of special surface action early on [Goobar et al., 2014]. That may teach us that the *standard* explosions of this type may *appear* as rather spherically symmetric unfoldings of the debris (which re-radiates the radioactive energy input, and may thus have a smoothing effect as a bolometer); but the explosion itself does not necessarily also have to reflect the same spherically symmetric properties, and may in fact allow more complex ignition processes to be involved [Röpke et al., 2011].

In a way, this is also hinted at by high resolution spectroscopy of the emerging ^{56}Co decay lines (see Fig. 4): The appearance is less smooth and gradual than would be expected by a smooth unfolding of a spherically symmetric envelope including the inner ^{56}Ni ashes [Diehl et al., 2015]. Gamma rays at energies around the ^{56}Co laboratory values appear to come at different times, to gradually build up the Doppler broadened 40 keV wide lines seen at brightness maximum. But the light curve rises more slowly than standard models suggest, and indicate irregularity that points to a heterogeneous distribution of ^{56}Ni throughout the expanding supernova. Thus, while being an impressive confirmation of our general understanding of thermonuclear supernovae, the

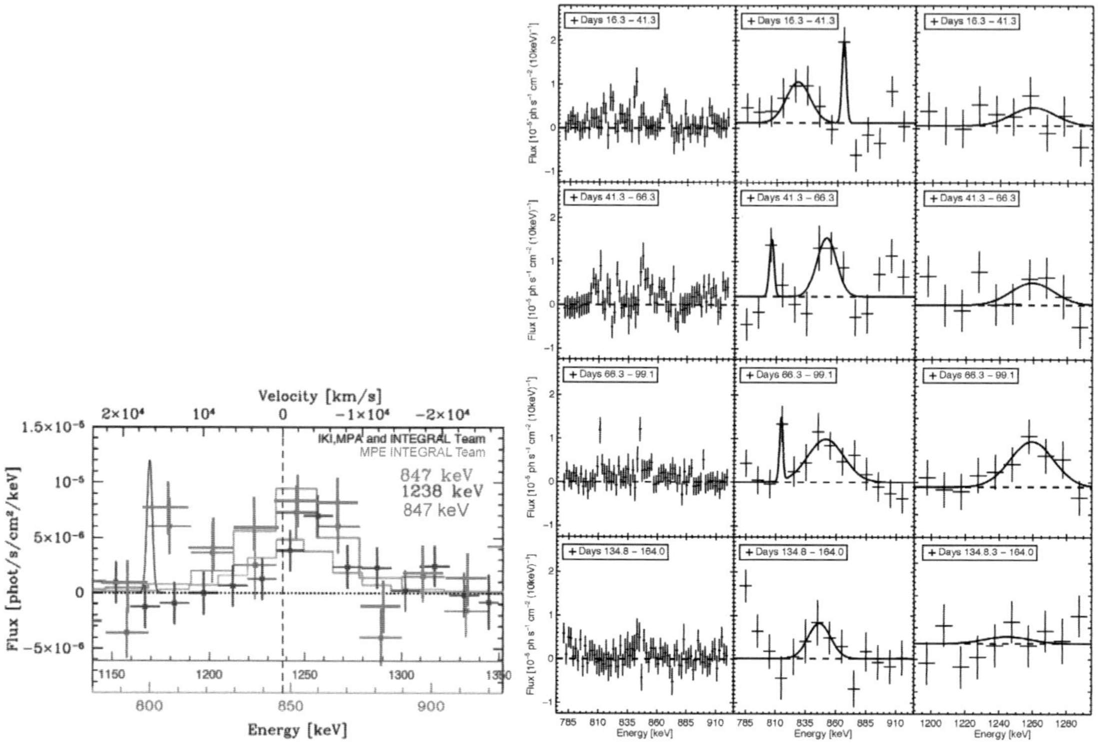

Figure 4. The 847 keV line from ^{56}Co decay from SN2014J measured with INTEGRAL/SPI appears broad, and consistently found from two different teams if analysed in broad energy bins [Churazov et al., 2014; Diehl et al., 2015] (*left*). The line intensity appears to rise irregularly, if analysed in fine and broad energy bins and for different phases during the increasing transparency of the supernova envelope [Diehl et al., 2015] (*right*).

gamma-ray data also present some interesting hints and puzzles upon a closer look. Similar to the SN1987A event, which changed how we thought core-collapse supernovae to arise, SN2014J has the potential to lead to major refinements and changes to our supernova type Ia models.

With respect to core collapse supernovae, SN1987A has been teaching us new aspects ever since the explosion almost thirty years ago, as we can follow the phenomena which arise and evolve as we are able to observe them at just 50 kpc distance. Recently, the NuSTAR hard X-ray mirror telescope has been able to measure the low-energy line at 68 keV from ^{44}Ti decay with great precision [Boggs et al., 2015]. This confirms the earlier INTEGRAL detection of these lines [Grebenev et al., 2012] and thus the direct proof of ^{44}Ti decay that had been inferred indirectly only from the late, shallow, decline of the bolometric emission of the young supernova remnant. The NuSTAR measurement reports a lower intensity than what INTEGRAL had reported, and the 1.5 10^{-4} $M_{\odot}$ of ^{44}Ti which are inferred are still higher than standard model yields, but consistent with the late-lightcurve inferred values. The slight red-shift of the ^{44}Ti line also confirms other indications that the inner explosion had some deviations from spherical symmetry.

The NuSTAR image in ^{44}Ti 68 keV emission [Grefenstette et al., 2014] (Fig. 5) is a great achievement for nuclear astrophysics. It clearly demonstrates that such deviations from spherically-symmetric explosions, and clumpy structures in the inner ejecta, are an observational fact. Moreover, the fact that the Chandra X-ray image showing Fe as seen in recombination

Journal of Physics: Conference Series **703** (2016) 012001 doi:10.1088/1742-6596/703/1/012001

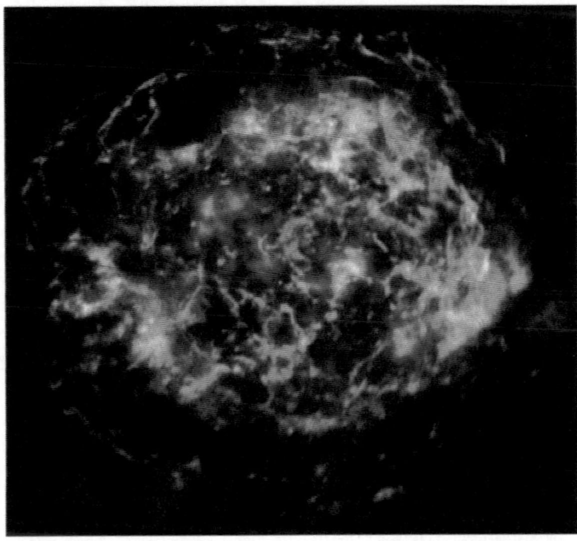

Figure 5. The ^{44}Ti decay image of Cas A, as measured with NuSTAR from the hard X-ray lines of ^{44}Sc de-excitation. This image resolves the emission regions, and shows irregular and clumpy ejecta (from Grefenstette et al. [2014])

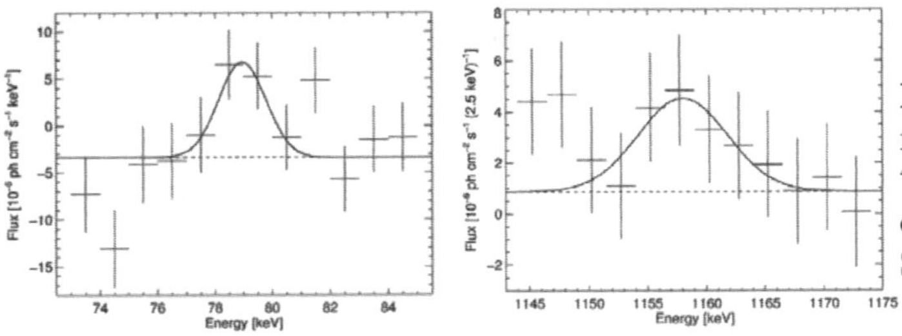

Figure 6. The lines from ^{44}Ti decay from Cas A, from de-excitation of ^{44}Sc (*left*), and from de-excitation of ^{44}Ca (*right*) as measured with SPI/INTEGRAL. (from Siegert et al. [2015])

lines differs in detail from the NuSTAR image also is interesting: Explosion physics strongly suggests that Ti and Fe both arise from th very inner supernova regions, and thus should be co-located; it is difficult to imagine a process that would be able to separate them as much as these image comparisons suggest. Rather, it seems plausible that the X-ray recombination emission includes a bias from the ionisation of Fe, which thus does not reflect *all of* the Fe in the expanding remnant. The Fe that is co-located with the ^{44}Ti emission clumps apparently is not X-ray bright on average.

INTEGRAL's SPI spectrometer recently also added new insight, from re-analysis of archival data [Siegert et al., 2015]. Now the lines from ^{44}Ti decay at low energy (68 and 78 keV lines originate from the first decay stage and from de-excitation of ^{44}Sc) as well as the line at 1157 keV originating from de-excitation of ^{44}Ca have both been measured by the same instrument (Fig. 6). This is important, in view of instrumental systematics, which always is a concern for measurements in the MeV regime and with instruments that suffer from a large instrumental background. Both lines show large Doppler broadening (4300$\pm$1600 / 2200$\pm$1600 km s^{-1} for the 78 / 1157 keV line, respectively), consistent with the expected ejecta expansion velocity of several 1000 km s^{-1}. The high-energy line appears somewhat brighter than the low-energy line; this brightness ratio also results from comparing all measurements of ^{44}Ti decay in Cas A since the pioneering COMPTEL detection of the 1157 keV line [Iyudin et al., 1994], and the various X-ray line measurements thereafter (Fig. 7) [Siegert et al., 2015; Vink, 2004] . Since the brighter line originates from ^{44}Ca that does not decay any further, it may be possible that nuclear excitation from locally-produced cosmic rays may add to the brightness of the 1157 keV line. Note that young supernova remnants are believed to be the main sources of cosmic rays in the galaxy [Blasi, 2013]; but direct proof has not been possible, as the leakage out of a

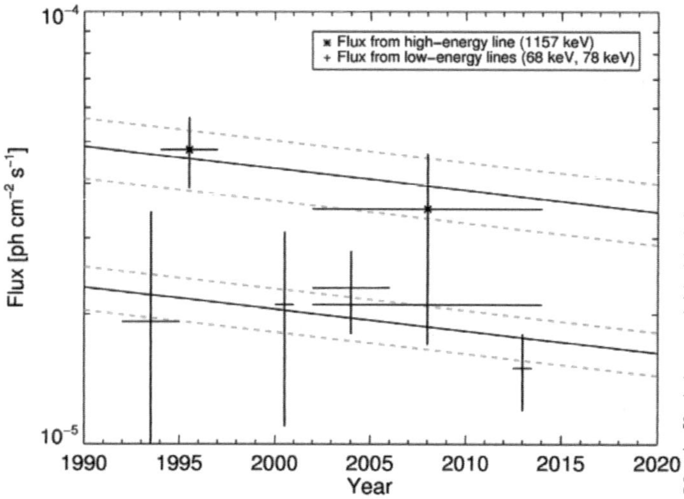

Figure 7. The lines from ^{44}Ti decay from Cas A, as measured with different instruments (data points with error bars), reflect the decay time of ^{44}Ti (shown as solid lines). The 1157 keV line from de-excitation of ^{44}Ca appears somewhat brighter, which could hint towards particle acceleration. (from Siegert et al. [2015])

supernova remnant is expected to occur after typically 1000 years only, when the outer shock weakens, and hence when X- and radio emission from these outer shock regions are difficult to detect. Fermi and H.E.S.S./MAGIC have made measurements of the impacts of cosmic-ray leakage onto the surroundings, in particular molecular clouds, to demonstrate cosmic ray acceleration in supernova remnants; but the difficulty to spectrally distinguish between inverse-Compton acceleration from cosmic ray electrons and pions from cosmic-ray protons has obtained different conclusions for different sources. Low energy cosmic ray interactions such as suggested by this brightness ratio measurement may be searched for, to test this cosmic ray acceleration in a young supernova remnant.

3.2. Diffuse nucleosynthesis and positrons

If radioactive decays occur at time scales that are long with respect to the re-occurrence time of nucleosynthesis events, the contributions from individual events will overlap and be super-imposed. Considering that for our Galaxy, the rate of supernovae of type core-collapse is about one supernova every 50 years[Diehl et al., 2006a], and the rate of novae is about 30-35 events per year, it is clear that all radioactivities beyond the 100 year scale will be of diffuse nature. For the study of such diffuse emission, therefore, it is required that the interpretation accounts for superposition from many sources. The spatial distribution of candidate sources, i.e. their distances and locations within the Galaxy, and also the potential variety in their nucleosynthesis yields have to be modelled from prior knowledge, which can be theory or resulting from other astronomical constraints or combinations thereof. Then, such a source model must be integrated over the radioactive decay time, accounting for temporal evolution of the rate of events, or else, one must adopt a steady state. The latter is plausible and often adopted, i.e. for time scales of My, which relates to the long-lived gamma ray emission sources of ^{26}Al and ^{60}Fe and also positron annihilation, one may assume that the event rate in our *current* Galaxy did not change over the past few My, and a steady state has been obtained from the new production and the radioactive decay of unstable nuclei resulting from cosmic nucleosynthesis.

3.2.1. ^{26}*Al* The discovery of characteristic gamma rays from the radioactive decay to ^{26}Al [Mahoney et al., 1982] from the extended inner region of our Galaxy was proof that the synthesis of this isotope, and hence cosmic nucleosynthesis, was ongoing in our Galaxy. This was interpreted as support for above steady-state hypothesis, rather than attributing such ^{26}Al

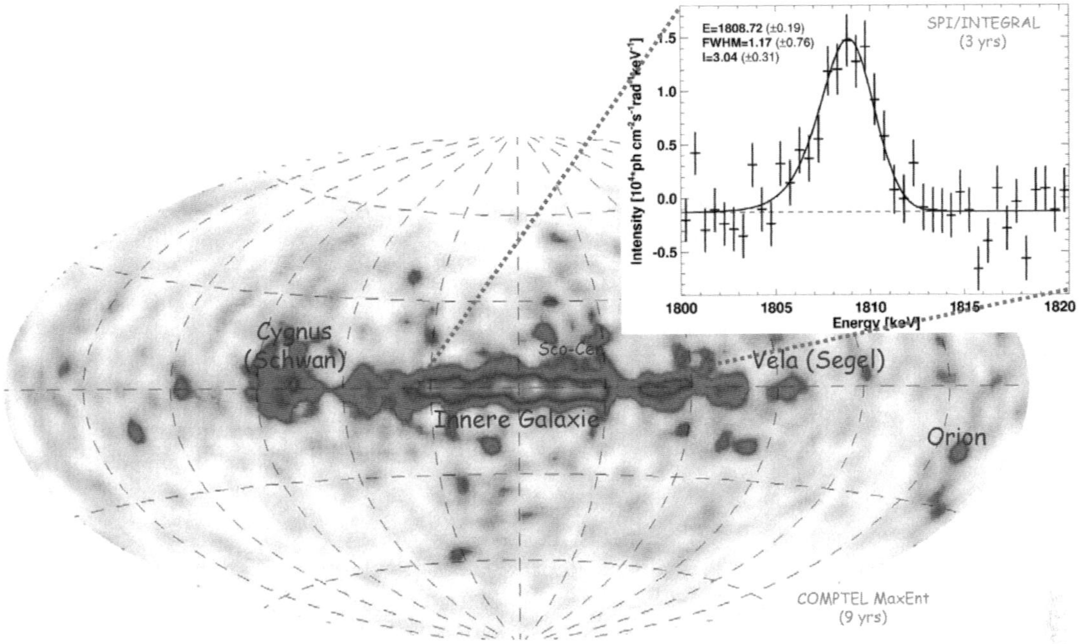

Figure 8. The ^{26}Al image derived with COMPTEL [Plüschke et al., 2001], and a spectrum from the inner Galaxy as measured with INTEGRAL/SPI [Diehl et al., 2006b]. This line reflects current nucleosynthesis in the Galaxy over the past Myrs. Its analysis has evolved into a rich field of astrophysics of the structure of our Galaxy, as well as of massive star groups (see text).

to a peculiar recent and nearby event (which would have to be located towards the direction of the inner Galaxy, coincidently). Since then, gamma ray surveys have confirmed that ^{26}Al gamma rays originated from diffuse emission across large regions of the sky, along the plane of our Galaxy. In particular, the COMPTEL data (Fig. 8) collected in years 1991-2000 of the CGRO Mission provided support for a sky image that identified the inner Galactic ridge and a few nearby massive star regions as sources of ^{26}Al. With INTEGRAL's spectrometer, since 2002 data are being taken on the spectral characteristics of this ^{26}Al emission (Fig. 8). This has led to a re-determination of the total content of ^{26}Al in our Galaxy under the steady-state assumption, and also to the detailed comparison of nucleosynthetic production models considering the source content of localised regions with their population of (mainly massive) stars. Moreover, the line shape measurements also set kinematic constraints to the ^{26}Al source regions. We will briefly discuss and summarise these aspects in the following.

The Galactic ^{26}Al content. The determination of the total mass of ^{26}Al contained in our Galaxy can be used for a global comparison to models: If one integrates over the different candidate sources, accounting for their yields of interstellar ^{26}Al and their spatial distribution (distances), one obtains a general consistency check of our understanding as reflected in models with observational constraints. Early such comparisons [Clayton and Leising, 1987; Signore and Dupraz, 1993] based on data from HEAO-C [Mahoney et al., 1982] and SMM-GRS [Share et al., 1985] thus saw novae, AGB stars, Wolf-Rayet stars, and core-collapse supernovae as plausible sources, with sufficient uncertainties in yields to regards each of these as significant contributors. The total content of interstellar Galactic ^{26}Al was determined as 2–3 $M_\odot$. With COMPTEL's image of the ^{26}Al emission across the entire sky [Diehl et al., 1995; Oberlack et al., 1996; Plüschke et al., 2001], it became clear [Prantzos and Diehl, 1996] that massive stars, i.e. core-collapse

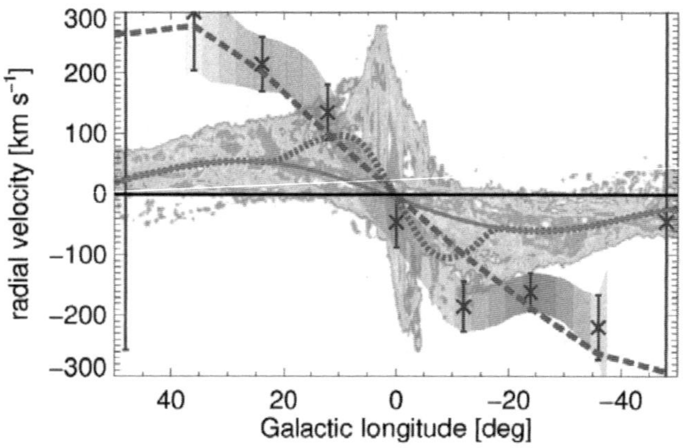

Figure 9. The ^{26}Al line shows a clear Doppler shift from large-scale rotation of the Galaxy. The velocities, however, are systematically higher by about 200 km s^{-1}. This is interpreted as ^{26}Al being ejected into superbubbles that have expanded more towards the direction forward and away from spiral arms, as the ejection occurs after the groups have evolved over several Myrs and moved away from their formation sites. (From Kretschmer et al. [2013]).

supernovae and Wolf-Rayet stars, were dominant contributors, an argument mainly based on the irregular and clumpy appearance of the ^{26}Al gamma-ray emission along the plane of the Galaxy, pointing to prominent contributions from clusters of massive stars. These comparisons made use of the bright emission along the inner Galactic ridge, fitting this with intensity-scaled models of spatial source distributions such as disk models based on exponential fall-off with a characteristic scale radius (about 2-4 kpc) and scale height (about 100-300 pc); but also more complex models such as spiral arm representations were used. It became clear, however, that a few more nearby source regions, such as the prominent Cygnus region, or in particular the Scorpius-Centaurus groups at only 110-150 pc distance, could lead to a systematic over-estimate of the Galactic mass of ^{26}Al, if not properly accounted for [Diehl et al., 2010; Martin et al., 2009].

^{26}Al and massive star groups. Predicting the contributions to interstellar ^{26}Al from a group of massive stars turned out to be crucial for any bottom-up comparison of source yields with gamma-ray data. Here, a model-predicted yield for a star of mass M must be multiplied with the number of occurrences of stars of this mass, $N(M)$. Then, stellar evolution must be accounted for, as stellar wind phases release ^{26}Al much earlier than the core-collapse supernova, which is relevant considering the decay time of ^{26}Al of 1.40 My. Population synthesis models were built [Cerviño et al., 2002; Martin et al., 2010; Voss et al., 2009] to put such predictions on a firm footing for comparisons to a range of astronomical messengers, including gamma-ray data, but also interstellar ionisation, interstellar cavity sizes, and stellar emission at other wavelengths. These models mostly used a mass distribution of stars from Salpeter's law, $dN(M)/dM \propto M^{-2.3}$; see however discussion of the present-day mass function aspects [Weidner and Kroupa, 2005; Weidner et al., 2010]. Comparisons were made for specific massive-star groups, where astronomical constraints were available on the stellar census, on interstellar cavities, and on ^{26}Al emission. This led to corrections of the massive star content in the Cygnus region [Knödlseder, 2000; Martin et al., 2009], to hints for wind model adjustments from the Carina region [Voss et al., 2012], and also provided a basis for more detailed investigations for the nearby Orion [Diehl, 2002; Diehl et al., 2003; Fierlinger et al., 2012, 2016; Voss et al., 2010] and Scorpius-Centaurus [Diehl et al., 2010; Pöppel et al., 2010; Preibisch et al., 2002] groups, where astronomical constraints are considered rather complete. These studies therefore are ongoing, and allow, a.o., for testing stellar evolution models with and without inclusion of stellar rotation, for different mass functions including non-explosion of some mass intervals in the 10–30 $M_\odot$ range, stellar wind and supernova overlaps in superbubbles, and correlated star formation scenarios (triggered star formation).

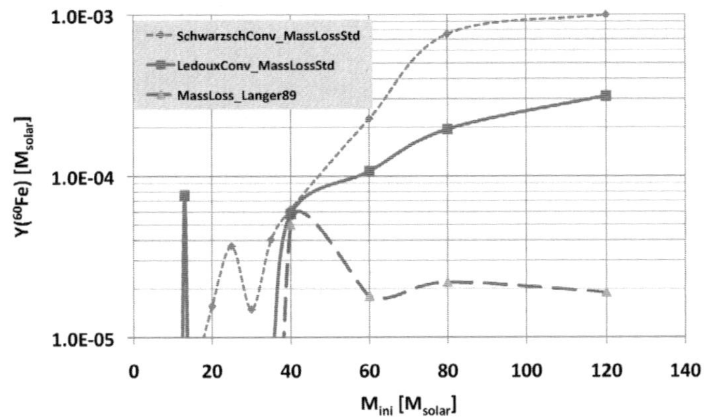

Figure 10. The yields in ^{60}Fe from massive stars are rather uncertain. In this graph, adapted from Limongi and Chieffi [2006b], we illustrate the impact of different treatments of known physical processes inside the star.

Implications of the Galactic ^{26}Al line characteristics. INTEGRAL's spectrometer has been maintaining the excellent spectral resolution of its Ge detectors over more than 13 mission years now, through periodic annealing operations of the detectors. Here, roughly every 6 months, the cryo-cooled Ge detector array with its 19 elements is elevated from normal operation temperatures (~80 K) to ~10° C for about two weeks, thus curing the lattice defects in Ge detectors which compromise charge collection, and are created by cosmic-ray bombardments in space. As a result, the survey data of INTEGRAL over many years can be analysed coherently at high spectral resolution. This allowed to recognise the large-scale Galactic rotation in the ^{26}Al line through its characteristic systematic Doppler shifts with Galactic longitude viewing direction (Fig. 9) [Diehl et al., 2006a; Kretschmer et al., 2013]. These measurements led to a precision of line shape constraints, which allowed to compare the large-scale Galactic rotation velocities with longitude between ^{26}Al, which traces nucleosynthesis ejecta, and CO, which traces molecular clouds parental to new star formation and massive star groups [Kretschmer et al., 2013]. The apparent discrepancy of velocities, with an excess velocity of ~200 km s^{-1} of ^{26}Al-based values over CO-based values, came as a surprise. The explanation was obtained from modelling how massive star groups form, evolve, and dissolve over tens of My, recognising that the ^{26}Al ejection mostly will occur into cavities created from stellar winds and earlier star formation, and that such cavities are more likely to occur between spiral arms than within them. Therefore, superbubbles are recognised [Krause et al., 2015] through these ^{26}Al measurements as major elements in the process of returning nucleosynthesis ejecta from stars and supernovae into cold interstellar gas that may later form next-generation stars. We have tested several aspects of this *stellar feedback* with respect to X-ray emission [Krause et al., 2014] and the origins of globular clusters with their peculiar abundance patterns [Krause et al., 2016]. At the galactic scale, superbubbles will also influence how the halo and interstellar clouds herein are involved into the cycling of matter from star forming region to next-generation star forming region [Fraternali, 2012], and how disk and halo exchange angular momentum [Krause et al., 2015; Kretschmer et al., 2013].

3.2.2. ^{60}Fe The gamma-ray emission from ^{60}Fe was predicted from nucleosynthesis theory [Clayton, 1971]. This was re-affirmed after ^{26}Al has been found, and the massive star source hypothesis turned out to be more plausible and dominant as compared to a nova origin [Timmes et al., 1995]. Massive star interiors are candidate sites of neutron capture on iron nuclei and thus ^{60}Fe producers, in particular in the He and C shells where neutron release from the ^{13}C(α,n) and ^{22}Ne(α,n) reactions were expected [Limongi and Chieffi, 2006a]. It just remained unclear how much of the ^{60}Fe would survive and not be destroyed by further neutron captures. Predictions

Journal of Physics: Conference Series **703** (2016) 012001 doi:10.1088/1742-6596/703/1/012001

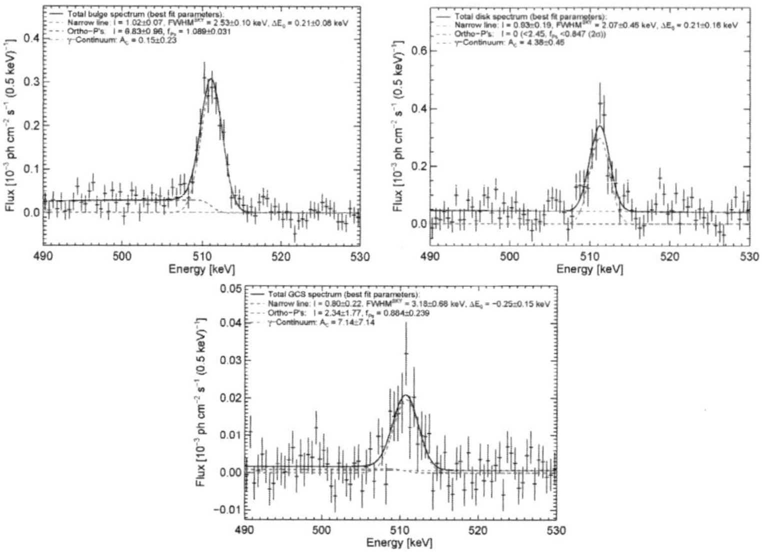

Figure 11. The spectra obtained from separating contributions from the bright bulge (*top left*), the extended disk (*top right*), and a potential point source in the central Galaxy (*bottom*) (from Siegert et al. [2016]).

range from insignificant amounts to amounts that rival ^{26}Al yields, yet strongly depend on uncertainties of massive star shell burning astrophysics as well as the neutron capture and beta decay reaction rates involved (see Fig. 10). The shell burning stages within massive stars in their late evolution are quite uncertain: Here, 3D effects of gas motions from shellular rotation stimulated by the rotating star play a key role, as well

Around 2005, gamma rays from ^{60}Fe decays in the Galaxy were reported from both the RHESSI [Harris et al., 2005] and the INTEGRAL [Harris et al., 2004] observations of our Galaxy. The early reports were not highly significant detections, but after more INTEGRAL observations seemed consolidated [Wang et al., 2007]. It became clear that the gamma ray brightness of ^{60}Fe was far below the brightness in ^{26}Al, the intensity ratio being about 15% ($\pm$5%). Note that β-decay lifetime measurements as well as neutron capture measurements are far from simple, and recent efforts have contributed here [Heftrich et al., 2015; Rugel et al., 2009]. This allows some interesting astrophysical constraints, yet precludes a detailed study of line shapes and regional/localised emission regions, which had turned ^{26}Al studies into such a rich field in the meantime.

3.2.3. Positron annihilation Positron emission results from the production of proton rich isotopes, in general. This occurs mainly in hydrogen burning in stars and nova explosions. Thus positron annihilation may be plausibly expected to occur after nova explosions, but also in type-I X-ray bursting sources in neutron star binaries; in normal stars, nucleosynthesis release times into interstellar space are longer than any of the decay times of β^+ radioactivities. Therefore, the study of positron annihilation across our Galaxy also is related to the field of nuclear astrophysics, and the contribution to positrons from nucleosynthesis have been studied since early times [Clayton, 1973; Lingenfelter and Ramaty, 1978; Ramaty et al., 1994]. A recent assessment [Martin et al., 2012], including the propagation of positrons from their nucleosynthesis sources to potential annihilation sites, has updated these expected contributions to positron annihilation gamma rays from nucleosynthesis sources.

Beyond the connection to nucleosynthesis and β^+-unstable isotopes, positrons may also be produced from other objects, such as pair plasma creating high-energy sources (pulsars, accreting binaries), and by cosmic ray interactions in interstellar space [Prantzos et al., 2011].

Already from OSSE results from the Compton Gamma Ray Observatory mission, it had been suggested that the annihilation gamma-ray emission was diffuse in nature, but particularly bright in the region of the inner Galaxy and its bulge region [Purcell et al., 1997]. These data had ruled out the transient annihilation hypothesis based on the 'Great Annihilator' results discussed earlier [Leventhal et al., 1989; Mirabel et al., 1992; Ramaty et al., 1992]. INTEGRAL observations then have provided a rather detailed picture from the gamma ray sky in positron annihilation emission[Bouchet et al., 2010; Churazov et al., 2005; Jean et al., 2003, 2006; Knödlseder et al., 2005; Skinner et al., 2012; Weidenspointner et al., 2008]. It now was clear that the inner Galaxy with its bulge region was a very prominent source of annihilation gamma rays, much brighter than the disk region throughout the Galaxy itself. This was surprising, as practically all candidate sources would be expected to populate the Galactic disk . Also surprising was the rather regular appearance of the central Galaxy emission, slightly offset from longitude zero by about 1 degree, but otherwise rather smooth in distribution and circularly extended both in longitude and latitude, suggesting an extent of 1–2 kpc. This stimulated interpretations of dark matter as a main source, as in this case the Galaxy's gravitational potential would dictate the spatial distribution of the emission. However, the ^{26}Al emission along the plane of the Galaxy demonstrates that positron sources in the Galactic disk regions do exist, and should contribute to the gamma-ray emission.

After more than four years of data collection, INTEGRAL began to reveal the disk emission in positron annihilation gamma-rays as well [Bouchet et al., 2010; Weidenspointner et al., 2008]. Some discussion about asymmetries ensued [Weidenspointner et al., 2008], and was later understood as originating from the above-mentioned slight offset of the emission centroid [Bouchet et al., 2010], while analyses searched for differences in brightness at positive versus negative longitudes throughout the Galaxy [Skinner et al., 2012]. Recently, from 11 years of data, the spatial distribution and spectroscopy has been re-assessed (Fig. 11) [Siegert et al., 2016]. In those latest results it became clear that the disk also is a substantial emitter of positron annihilation gamma rays, rivalling the bulge region in total brightness, but with much lower surface brightness. Therefore it had been difficult to detect this disk emission with less exposure in earlier data. Also, the disk appears now to be more extended in latitude than believed earlier, reducing the surface brightness even more. This suggests that positrons do escape from source regions also towards the halo, thus leading to increased latitude extent of the gamma ray emission. Possibly, positrons escape into the halo, and may then be re-directed into the inner Galaxy and bulge region by large-scale magnetic fields in the halo, following an earlier model to explain the bright bulge emission.

The annihilation of positrons may occur after slowing down from the relativistic energies at which they typically are produced, and only at energies in the eV region. Then, in interstellar space the formation of positronium through charge-exchange collisions with hydrogen is important, and an intermediate stage that eases annihilation because of angular momentum conservation. Positronium is an atom-like bound quantum mechanical system composed of the positron and an electron, with a binding energy of 6.8 eV. From its relative spin orientations, it can adopt a singlet and a triplet status. Annihilation then can occur from either of these states, the triplet state being somewhat more long-lived and producing three photons from angular momentum conservations, while the singlet state produces two photons of 511 keV energy. Annihilation thus converts the electron-positron rest masses of 1.022 MeV into radiation, producing a line at 511 keV and a characteristic continuum towards lower energies. The relative intensities of these two spectral components encode how much annihilation occurs through the positronium intermediate state, versus direct annihilation that is the only annihilation channel

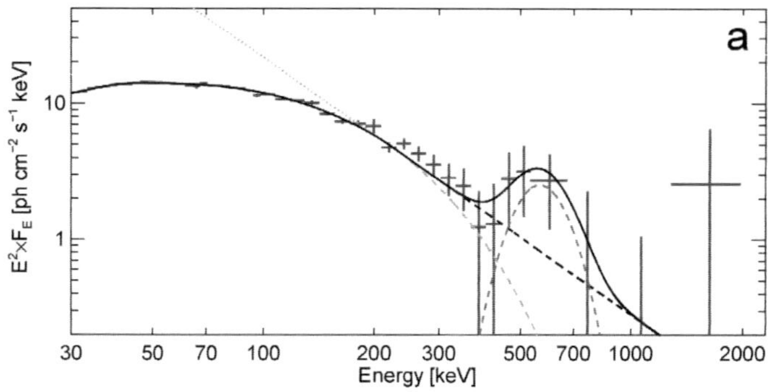

Figure 12. The spectrum of microquasar V404 Cygni in its early flaring period June 2015, obtained from INTEGRAL/SPI observations (from Siegert [2016]). Here, the annihilation occurs in hot plasma clouds near the black hole, kinematic shift and broadening from the Doppler effect results in the bumb feature between 0.5 and 1 MeV, consistent with theoretical expectations.

in fully-ionised interstellar plasma. From this, it had been deduced that positrons typically annihilate in a partially-ionised medium with sufficient neutral atoms to make positronium annihilation the dominant channel, and from the width of the 511 keV line the temperature of the medium is constrained through Doppler broadening to about 7000 K [Churazov et al., 2005, 2015]. In latest results [Siegert et al., 2016], the different spectral components and line width could be identified separately in emission attributed to the bright bulge region, to the extended and faint disk region, and to another point-like source near the Galactic Centre direction. It appears that spectra may differ between bulge and disk (Fig. 11, upper spectra) , indicating different typical/averaged annihilation conditions. Also, there are indications that the central source shows a broader 511 keV line (Fig. 11, below) , which may suggest that annihilations here occur in gas with higher kinematic motion or at higher temperature, such as expected if the central black hole environment was related to this source.

One of the candidate positron sources are accreting binary systems with accretion onto a stellar-mass black hole; these objects are called micro-quasars [Prantzos et al., 2011]. Here, the matter is channeled towards the black hole through an accretion disk that shines in X-rays [Done et al., 2007; Remillard and McClintock, 2006], and disappears in the black hole, somehow resulting in plasma ejections seen as jets of radio emission far out in those systems [Mirabel and Rodríguez, 1999]. Such binary system accretion often is unstable, reflecting different configurations of the accretion flow and the geometrical shapes of the accretion disk. One of these microquasars, V404 Cygni, nearby in our Galaxy, went into a flaring state in June 2015 [Rodriguez et al., 2015], thus becoming the brightest source in the X-ray sky for several weeks. Analysing INTEGRAL data from this flaring period, it was recently recognised [Siegert, 2016] that the typical gamma-ray signatures of positron annihilation were present (Fig. 12), although significantly Doppler broadened and variable, and thus not easy to detect, typically. This underlines the candidate source role of microquasars to produce positrons which later annihilate in interstellar space. The local annihilation seen recently with V404 Cygni occurs near the production site very close to the black hole itself, and could be seen in this source only because it is bright enough to also have sufficient brightness in such a side channel of its total radiation spectrum.

4. Summary and perspectives

The measurement of cosmic gamma rays in the nuclear domain, from ~100 keV to a few MeV, have evolved into a diverse and rather fruitful field of high energy astronomy. Its science is rather unique, addressing cosmic sources of nucleosynthesis, of particle acceleration, and of positron production. Particle acceleration can also be studied at higher energies in gamma rays, and with cosmic rays. The other fields are unique to this spectral window. Current instrumentation features Compton scattering or coded mask techniques for spatial resolution, and semiconductors for spectral precision. Through better segmentation of detectors and tracking of secondary particles, the precision and also sensitivity could be advanced now to explore sources beyond our Galaxy and many expected but fainter Galactic sources. But space missions are costly and sparse. INTEGRAL's spectrometer, together with the NuSTAR hard X-ray telescope, are the only current missions which provide measurements in this energy range. No further such instruments are currently planned, and therefore will probably not be realised within the next decades (space projects typically have 10-20 years of lead time). But INTEGRAL is working well after 13 years, even though it was originally planned for a 5-year maximum of operations. Its orbit and onboard resources should allow proper functionality until 2029, when the spacecraft will re-enter the Earth atmosphere and be de-orbited, unless some unforeseen malfunction or micrometeorite collision would occur earlier. Re-approval after science and operations review occurs every two years at ESA, and currently the mission is financed till end 2018. As above results show, important scientific refinements of gamma-ray results occur in the mature and late INTEGRAL mission, and transient sources such as SN2014J and V404 Cygni make these mission extensions more worthwhile.

Acknowledgments

The INTEGRAL/SPI project has been completed under the responsibility and leadership of CNES; we are grateful to ASI, CEA, CNES, DLR, ESA, INTA, NASA and OSTC for support of this ESA space science mission. R.D. acknowledges support from the Munich cluster of excellence *origin and evolution of the universe*, and from the organisers of this summer school.

References

D. L. Bertsch, C. E. Fichtel, and J. I. Trombka. Instrumentation for gamma-ray astronomy. *Space Sci. Rev.*, 48:113–168, Mar. 1988. doi: 10.1007/BF00183131.

P. Blasi. The origin of galactic cosmic rays. *Astron.&Astroph. Rev.*, 21:70, Nov. 2013. doi: 10.1007/s00159-013-0070-7.

S. E. Boggs, F. A. Harrison, H. Miyasaka, B. W. Grefenstette, A. Zoglauer, C. L. Fryer, S. P. Reynolds, D. M. Alexander, H. An, D. Barret, F. E. Christensen, W. W. Craig, K. Forster, P. Giommi, C. J. Hailey, A. Hornstrup, T. Kitaguchi, J. E. Koglin, K. K. Madsen, P. H. Mao, K. Mori, M. Perri, M. J. Pivovaroff, S. Puccetti, V. Rana, D. Stern, N. J. Westergaard, and W. W. Zhang. ^{44}Ti gamma-ray emission lines from SN1987A reveal an asymmetric explosion. *Science*, 348:670–671, May 2015. doi: 10.1126/science.aaa2259.

BOREXINO Collaboration, C. Arpesella, G. Bellini, J. Benziger, S. Bonetti, B. Caccianiga, F. Calaprice, F. Dalnoki-Veress, D. D'Angelo, H. de Kerret, A. Derbin, M. Deutsch, A. Etenko, K. Fomenko, R. Ford, D. Franco, B. Freudiger, C. Galbiati, S. Gazzana, M. Giammarchi, M. Goeger-Neff, A. Goretti, C. Grieb, S. Hardy, G. Heusser, A. Ianni, A. Ianni, M. Joyce, G. Korga, D. Kryn, M. Laubenstein, M. Leung, E. Litvinovich, P. Lombardi, L. Ludhova, I. Machulin, G. Manuzio, A. Martemianov, F. Masetti, K. McCarty, E. Meroni, L. Miramonti, M. Misiaszek, D. Montanari, M. E. Monzani, V. Muratova, L. Niedermeier, L. Oberauer, M. Obolensky, F. Ortica, M. Pallavicini, L. Papp, L. Perasso, A. Pocar, R. S. Raghavan, G. Ranucci, A. Razeto, A. Sabelnikov, C. Salvo, S. Schönert, H. Simgen, O. Smirnov,

M. Skorokhvatov, A. Sonnenschein, A. Sotnikov, S. Sukhotin, Y. Suvorov, V. Tarasenkov, R. Tartaglia, G. Testera, D. Vignaud, S. Vitale, R. B. Vogelaar, F. von Feilitzsch, M. Wojcik, O. Zaimidoroga, S. Zavatarelli, and G. Zuzel. First real time detection of ^{7}Be solar neutrinos by Borexino. *Physics Letters B*, 658:101–108, Jan. 2008. doi: 10.1016/j.physletb.2007.09.054.

BOREXINO Collaboration, G. Bellini, J. Benziger, D. Bick, G. Bonfini, D. Bravo, B. Caccianiga, L. Cadonati, F. Calaprice, A. Caminata, P. Cavalcante, A. Chavarria, A. Chepurnov, D. D'Angelo, S. Davini, A. Derbin, A. Empl, A. Etenko, K. Fomenko, D. Franco, F. Gabriele, C. Galbiati, S. Gazzana, C. Ghiano, M. Giammarchi, M. Göger-Neff, A. Goretti, M. Gromov, C. Hagner, E. Hungerford, A. Ianni, A. Ianni, V. Kobychev, D. Korablev, G. Korga, D. Kryn, M. Laubenstein, B. Lehnert, T. Lewke, E. Litvinovich, F. Lombardi, P. Lombardi, L. Ludhova, G. Lukyanchenko, I. Machulin, S. Manecki, W. Maneschg, S. Marcocci, Q. Meindl, E. Meroni, M. Meyer, L. Miramonti, M. Misiaszek, M. Montuschi, P. Mosteiro, V. Muratova, L. Oberauer, M. Obolensky, F. Ortica, K. Otis, M. Pallavicini, L. Papp, L. Perasso, A. Pocar, G. Ranucci, A. Razeto, A. Re, A. Romani, N. Rossi, R. Saldanha, C. Salvo, S. Schönert, H. Simgen, M. Skorokhvatov, O. Smirnov, A. Sotnikov, S. Sukhotin, Y. Suvorov, R. Tartaglia, G. Testera, D. Vignaud, R. B. Vogelaar, F. von Feilitzsch, H. Wang, J. Winter, M. Wojcik, A. Wright, M. Wurm, O. Zaimidoroga, S. Zavatarelli, K. Zuber, and G. Zuzel. Neutrinos from the primary proton-proton fusion process in the Sun. *Nature*, 512:383–386, Aug. 2014. doi: 10.1038/nature13702.

L. Bouchet, J. P. Roques, and E. Jourdain. On the Morphology of the Electron-Positron Annihilation Emission as Seen by SPI/INTEGRAL. *ApJ*, 720:1772–1780, Sept. 2010. doi: 10.1088/0004-637X/720/2/1772.

M. Cerviño, R. Diehl, K. Kretschmer, and S. Plüschke. Radioactive isotopes in star forming regions. *New Astronomy Review*, 46:541–545, July 2002. doi: 10.1016/S1387-6473(02)00198-7.

E. Churazov, R. Sunyaev, S. Sazonov, M. Revnivtsev, and D. Varshalovich. Positron annihilation spectrum from the Galactic Centre region observed by SPI/INTEGRAL. *Mon.Not.Royal Astr. Soc.*, 357:1377–1386, Mar. 2005. doi: 10.1111/j.1365-2966.2005.08757.x.

E. Churazov, R. Sunyaev, J. Isern, J. Knödlseder, P. Jean, F. Lebrun, N. Chugai, S. Grebenev, E. Bravo, S. Sazonov, and M. Renaud. Cobalt-56 γ-ray emission lines from the type Ia supernova 2014J. *Nature*, 512:406–408, Aug. 2014. doi: 10.1038/nature13672.

E. Churazov, R. Sunyaev, J. Isern, I. Bikmaev, E. Bravo, N. Chugai, S. Grebenev, P. Jean, J. Knödlseder, F. Lebrun, and E. Kuulkers. Gamma-rays from Type Ia Supernova SN2014J. *ApJ*, 812:62, Oct. 2015. doi: 10.1088/0004-637X/812/1/62.

D. D. Clayton. New Prospect for Gamma-Ray-Line Astronomy. *Nature*, 234:291–+, Dec. 1971. doi: 10.1038/234291a0.

D. D. Clayton. Galaxy-Positronium origin of 476 keV galactic feature. *Nature*, 244:137–+, Aug. 1973.

D. D. Clayton and M. D. Leising. ^{26}Al in the interstellar medium. *Phys. Rep.*, 144:1–50, 1987. doi: 10.1016/0370-1573(87)90099-8.

D. D. Clayton and L. R. Nittler. Astrophysics with Presolar Stardust. *ARA&A*, 42:39–78, Sept. 2004. doi: 10.1146/annurev.astro.42.053102.134022.

D. D. Clayton, S. A. Colgate, and G. J. Fishman. Gamma-Ray Lines from Young Supernova Remnants. *ApJ*, 155:75–+, Jan. 1969. doi: 10.1086/149849.

R. Diehl. ^{26}Al production in the Vela and Orion regions. *New Astronomy Review*, 46:547–552, July 2002. doi: 10.1016/S1387-6473(02)00199-9.

R. Diehl. Nuclear Astrophysics with Gamma-Ray Line Astronomy. In M. Busso, R. G. Pizzone, C. Rolfs, C. Spitaleri, and A. Tumino, editors, *EAS Publications Series*, volume 27 of *EAS Publications Series*, pages 83–102, 2007. doi: 10.1051/eas:2007146.

R. Diehl. Nuclear astrophysics lessons from INTEGRAL. *Reports on Progress in Physics*, 76 (2):026301, Feb. 2013a. doi: 10.1088/0034-4885/76/2/026301.

R. Diehl. Cosmic Gamma-Ray Spectroscopy. *ArXiv e-prints*, arXiv1307(1307.4198D):1–8, July 2013b.

R. Diehl, C. Dupraz, K. Bennett, H. Bloemen, W. Hermsen, J. Knoedlseder, G. Lichti, D. Morris, J. Ryan, V. Schoenfelder, H. Steinle, A. Strong, B. Swanenburg, M. Varendorff, and C. Winkler. COMPTEL observations of Galactic ^{26}Al emission. *Astron.&Astroph.*, 298: 445–+, June 1995.

R. Diehl, K. Kretschmer, S. Plüschke, M. Cerviño, and D. H. Hartmann. Gamma-rays from massive stars in Cygnus and Orion. In K. van der Hucht, A. Herrero, and C. Esteban, editors, *A Massive Star Odyssey: From Main Sequence to Supernova*, volume 212 of *IAU Symposium*, pages 706–+, 2003.

R. Diehl, H. Halloin, K. Kretschmer, G. G. Lichti, V. Schönfelder, A. W. Strong, A. von Kienlin, W. Wang, P. Jean, J. Knödlseder, J.-P. Roques, G. Weidenspointner, S. Schanne, D. H. Hartmann, C. Winkler, and C. Wunderer. Radioactive ^{26}Al from massive stars in the Galaxy. *Nature*, 439:45–47, Jan. 2006a. doi: 10.1038/nature04364.

R. Diehl, H. Halloin, K. Kretschmer, A. W. Strong, W. Wang, P. Jean, G. G. Lichti, J. Knödlseder, J.-P. Roques, S. Schanne, V. Schönfelder, A. von Kienlin, G. Weidenspointner, C. Winkler, and C. Wunderer. ^{26}Al in the inner Galaxy. Large-scale spectral characteristics derived with SPI/INTEGRAL. *Astron.&Astroph.*, 449:1025–1031, Apr. 2006b. doi: 10.1051/0004-6361:20054301.

R. Diehl, M. G. Lang, P. Martin, H. Ohlendorf, T. Preibisch, R. Voss, P. Jean, J. Roques, P. von Ballmoos, and W. Wang. Radioactive ^{26}Al from the Scorpius-Centaurus association. *Astron.&Astroph.*, 522:A51+, Nov. 2010. doi: 10.1051/0004-6361/201014302.

R. Diehl, D. H. Hartmann, and N. Prantzos. *Astronomy with Radioactivities*, volume 812 of *Lecture Notes in Physics, Berlin Springer Verlag*. Springer: Berlin, Heidelberg, 2011.

R. Diehl, T. Siegert, W. Hillebrandt, S. A. Grebenev, J. Greiner, M. Krause, M. Kromer, K. Maeda, F. Röpke, and S. Taubenberger. Early ^{56}Ni decay gamma rays from SN2014J suggest an unusual explosion. *Science*, 345:1162–1165, Sept. 2014. doi: 10.1126/science.1254738.

R. Diehl, T. Siegert, W. Hillebrandt, M. Krause, J. Greiner, K. Maeda, F. K. Röpke, S. A. Sim, W. Wang, and X. Zhang. SN2014J gamma rays from the ^{56}Ni decay chain. *Astron.&Astroph.*, 574:A72, Feb. 2015. doi: 10.1051/0004-6361/201424991.

C. Done, M. Gierliński, and A. Kubota. Modelling the behaviour of accretion flows in X-ray binaries. Everything you always wanted to know about accretion but were afraid to ask. *Astron.&Astroph. Rev.*, 15:1–66, Dec. 2007. doi: 10.1007/s00159-007-0006-1.

K. M. Fierlinger, A. Burkert, R. Diehl, C. Dobbs, D. H. Hartmann, M. Krause, E. Ntormousi, and R. Voss. Molecular Cloud Disruption and Chemical Enrichment of the ISM Caused by Massive Star Feedback. In R. Capuzzo-Dolcetta, M. Limongi, and A. Tornambè, editors, *Advances in Computational Astrophysics: Methods, Tools, and Outcome*, volume 453 of *Astronomical Society of the Pacific Conference Series*, page 25, July 2012.

K. M. Fierlinger, A. Burkert, E. Ntormousi, P. Fierlinger, M. Schartmann, A. Ballone, M. G. H. Krause, and R. Diehl. Stellar feedback efficiencies: supernovae versus stellar winds. *Mon.Not.Royal Astr. Soc.*, 456:710–730, Feb. 2016. doi: 10.1093/mnras/stv2699.

J. Fossey, B. Cooke, G. Pollack, M. Wilde, and T. Wright. Supernova 2014J in M82 = Psn J09554214+6940260. *Central Bureau Electronic Telegrams*, 3792:1, Jan. 2014.

F. Fraternali. Modelling the gas kinematics in disk galaxies. In M. A. de Avillez, editor, *EAS Publications Series*, volume 56 of *EAS Publications Series*, pages 355–362, Sept. 2012. doi: 10.1051/eas/1256058.

N. Gehrels, C. E. Fichtel, G. J. Fishman, J. D. Kurfess, and V. Schonfelder. The Compton Gamma-Ray Observatory. *Scientific American*, 269:68–+, Dec. 1993.

A. Goobar, J. Johansson, R. Amanullah, Y. Cao, D. A. Perley, M. M. Kasliwal, R. Ferretti, P. E. Nugent, C. Harris, A. Gal-Yam, E. O. Ofek, S. P. Tendulkar, M. Dennefeld, S. Valenti, I. Arcavi, D. P. K. Banerjee, V. Venkataraman, V. Joshi, N. M. Ashok, S. B. Cenko, R. F. Diaz, C. Fremling, A. Horesh, D. A. Howell, S. R. Kulkarni, S. Papadogiannakis, T. Petrushevska, D. Sand, J. Sollerman, V. Stanishev, J. S. Bloom, J. Surace, T. J. Dupuy, and M. C. Liu. The Rise of SN 2014J in the Nearby Galaxy M82. *ApJ*, 784:L12, Mar. 2014. doi: 10.1088/2041-8205/784/1/L12.

S. A. Grebenev, A. A. Lutovinov, S. Tsygankov, and C. Winkler. Hard-x-ray emission lines from the decay of ^{44}ti in the remnant of supernova 1987a. *Nature*, (tbd)((tbd)):(accepted for publication), 2012.

B. W. Grefenstette, F. A. Harrison, S. E. Boggs, S. P. Reynolds, C. L. Fryer, K. K. Madsen, D. R. Wik, A. Zoglauer, C. I. Ellinger, D. M. Alexander, H. An, D. Barret, F. E. Christensen, W. W. Craig, K. Forster, P. Giommi, C. J. Hailey, A. Hornstrup, V. M. Kaspi, T. Kitaguchi, J. E. Koglin, P. H. Mao, H. Miyasaka, K. Mori, M. Perri, M. J. Pivovaroff, S. Puccetti, V. Rana, D. Stern, N. J. Westergaard, and W. W. Zhang. Asymmetries in core-collapse supernovae from maps of radioactive ^{44}Ti in CassiopeiaA. *Nature*, 506:339–342, Feb. 2014. doi: 10.1038/nature12997.

M. J. Harris, J. Knodlseder, P. Jean, E. Cisana, G. Lichti, R. Diehl, K. Kretschmer, A. Von Kienlin, J.-P. Roques, S. Schanne, G. Weidenspointner, and C. Wunderer. Preliminary results of INTEGRAL/SPI measurements of radioactive ^{60}Fe in the Galaxy. In *Bulletin of the American Astronomical Society*, volume 36 of *Bulletin of the American Astronomical Society*, pages 950–+, Aug. 2004.

M. J. Harris, J. Knödlseder, P. Jean, E. Cisana, R. Diehl, G. G. Lichti, J.-P. Roques, S. Schanne, and G. Weidenspointner. Detection of γ-ray lines from interstellar ^{60}Fe by the high resolution spectrometer SPI. *Astron.&Astroph.*, 433:L49–L52, Apr. 2005. doi: 10.1051/0004-6361:200500093.

F. A. Harrison, W. W. Craig, F. E. Christensen, C. J. Hailey, W. W. Zhang, S. E. Boggs, D. Stern, W. R. Cook, K. Forster, P. Giommi, B. W. Grefenstette, Y. Kim, T. Kitaguchi, J. E. Koglin, K. K. Madsen, P. H. Mao, H. Miyasaka, K. Mori, M. Perri, M. J. Pivovaroff, S. Puccetti, V. R. Rana, N. J. Westergaard, J. Willis, A. Zoglauer, H. An, M. Bachetti, N. M. Barrière, E. C. Bellm, V. Bhalerao, N. F. Brejnholt, F. Fuerst, C. C. Liebe, C. B. Markwardt, M. Nynka, J. K. Vogel, D. J. Walton, D. R. Wik, D. M. Alexander, L. R. Cominsky, A. E. Hornschemeier, A. Hornstrup, V. M. Kaspi, G. M. Madejski, G. Matt, S. Molendi, D. M. Smith, J. A. Tomsick, M. Ajello, D. R. Ballantyne, M. Baloković, D. Barret, F. E. Bauer, R. D. Blandford, W. Niel Brandt, L. W. Brenneman, J. Chiang, D. Chakrabarty, J. Chenevez, A. Comastri, F. Dufour, M. Elvis, A. C. Fabian, D. Farrah, C. L. Fryer, E. V. Gotthelf, J. E. Grindlay, D. J. Helfand, R. Krivonos, D. L. Meier, J. M. Miller, L. Natalucci, P. Ogle, E. O. Ofek, A. Ptak, S. P. Reynolds, J. R. Rigby, G. Tagliaferri, S. E. Thorsett, E. Treister, and C. M. Urry. The Nuclear Spectroscopic Telescope Array (NuSTAR) High-energy X-Ray Mission. *ApJ*, 770:103, June 2013. doi: 10.1088/0004-637X/770/2/103.

T. Heftrich, M. Bichler, R. Dressler, K. Eberhardt, A. Endres, J. Glorius, K. Göbel, G. Hampel, M. Heftrich, F. Käppeler, C. Lederer, M. Mikorski, R. Plag, R. Reifarth, C. Stieghorst, S. Schmidt, D. Schumann, Z. Slavkovská, K. Sonnabend, A. Wallner, M. Weigand, N. Wiehl, and S. Zauner. Thermal neutron capture cross section of the radioactive isotope ^{60}Fe. *Phys. Rev. C*, 92(1):015806, July 2015. doi: 10.1103/PhysRevC.92.015806.

W. Hillebrandt, M. Kromer, F. K. Röpke, and A. J. Ruiter. Towards an understanding of Type Ia supernovae from a synthesis of theory and observations. *Frontiers of Physics*, 8:116–143, Apr. 2013. doi: 10.1007/s11467-013-0303-2.

A. F. Iyudin, R. Diehl, H. Bloemen, W. Hermsen, G. G. Lichti, D. Morris, J. Ryan, V. Schoenfelder, H. Steinle, M. Varendorff, C. de Vries, and C. Winkler. COMPTEL observations of Ti-44 gamma-ray line emission from CAS A. *Astron.&Astroph.*, 284:L1–L4, Apr. 1994.

P. Jean, J. Knödlseder, V. Lonjou, M. Allain, J.-P. Roques, G. K. Skinner, B. J. Teegarden, G. Vedrenne, P. von Ballmoos, B. Cordier, P. Caraveo, R. Diehl, P. Durouchoux, P. Mandrou, J. Matteson, N. Gehrels, V. Schönfelder, A. W. Strong, P. Ubertini, G. Weidenspointner, and C. Winkler. Early SPI/INTEGRAL measurements of 511 keV line emission from the 4th quadrant of the Galaxy. *Astron.&Astroph.*, 407:L55–L58, Aug. 2003. doi: 10.1051/0004-6361:20031056.

P. Jean, J. Knödlseder, W. Gillard, N. Guessoum, K. Ferrière, A. Marcowith, V. Lonjou, and J. P. Roques. Spectral analysis of the Galactic positron annihilation emission. *Astron.&Astroph.*, 445:579–589, Jan. 2006. doi: 10.1051/0004-6361:20053765.

S. Jones, R. Hirschi, M. Pignatari, A. Heger, C. Georgy, N. Nishimura, C. Fryer, and F. Herwig. Code dependencies of pre-supernova evolution and nucleosynthesis in massive stars: evolution to the end of core helium burning. *Mon.Not.Royal Astr. Soc.*, 447:3115–3129, Mar. 2015. doi: 10.1093/mnras/stu2657.

J. Knödlseder. Cygnus OB2 - a young globular cluster in the Milky Way. *Astron.&Astroph.*, 360:539–548, Aug. 2000.

J. Knödlseder, P. Jean, V. Lonjou, G. Weidenspointner, N. Guessoum, W. Gillard, G. Skinner, P. von Ballmoos, G. Vedrenne, J.-P. Roques, S. Schanne, B. Teegarden, V. Schönfelder, and C. Winkler. The all-sky distribution of 511 keV electron-positron annihilation emission. *Astron.&Astroph.*, 441:513–532, Oct. 2005. doi: 10.1051/0004-6361:20042063.

M. Krause, R. Diehl, H. Böhringer, M. Freyberg, and D. Lubos. Feedback by massive stars and the emergence of superbubbles. II. X-ray properties. *Astron.&Astroph.*, 566:A94, June 2014. doi: 10.1051/0004-6361/201423871.

M. G. H. Krause, R. Diehl, Y. Bagetakos, E. Brinks, A. Burkert, O. Gerhard, J. Greiner, K. Kretschmer, and T. Siegert. ^{26}Al kinematics: superbubbles following the spiral arms?. Constraints from the statistics of star clusters and HI supershells. *Astron.&Astroph.*, 578: A113, June 2015. doi: 10.1051/0004-6361/201525847.

M. G. H. Krause, C. Charbonnel, N. Bastian, and R. Diehl. Gas expulsion in massive star clusters?. Constraints from observations of young and gas-free objects. *Astron.&Astroph.*, 587:A53, Mar. 2016. doi: 10.1051/0004-6361/201526685.

K. Kretschmer, R. Diehl, M. Krause, A. Burkert, K. Fierlinger, O. Gerhard, J. Greiner, and W. Wang. Kinematics of massive star ejecta in the Milky Way as traced by ^{26}Al. *Astron.&Astroph.*, 559:A99, Nov. 2013. doi: 10.1051/0004-6361/201322563.

M. Leventhal, C. J. MacCallum, S. D. Barthelmy, N. Gehrels, B. J. Teegarden, and J. Tueller. Reappearance of the annihilation line source at the Galactic Centre. *Nature*, 339:36–38, May 1989. doi: 10.1038/339036a0.

M. Limongi and A. Chieffi. Nucleosynthesis of ^{60}Fe in massive stars. *New Astronomy Review*, 50:474–476, Oct. 2006a. doi: 10.1016/j.newar.2006.06.005.

M. Limongi and A. Chieffi. Nucleosynthesis of ^{60}Fe in massive stars. *New Astron. Rev.*, 50: 474–476, Oct. 2006b. doi: 10.1016/j.newar.2006.06.005.

R. E. Lingenfelter and R. Ramaty. Gamma-ray lines - A new window to the Universe. *Physics Today*, 31:40–47, Mar. 1978.

W. A. Mahoney, J. C. Ling, A. S. Jacobson, and R. E. Lingenfelter. Diffuse galactic gamma-ray line emission from nucleosynthetic Fe-60, Al-26, and Na-22 - Preliminary limits from HEAO 3. *ApJ*, 262:742–+, Nov. 1982. doi: 10.1086/160469.

P. Martin, J. Knödlseder, R. Diehl, and G. Meynet. New estimates of the gamma-ray line emission of the Cygnus region from INTEGRAL/SPI observations. *Astron.&Astroph.*, 506: 703–710, Nov. 2009. doi: 10.1051/0004-6361/200912178.

P. Martin, J. Knödlseder, G. Meynet, and R. Diehl. Predicted gamma-ray line emission from the Cygnus complex. *Astron.&Astroph.*, 511:A86+, Feb. 2010. doi: 10.1051/0004-6361/200912864.

P. Martin, A. W. Strong, P. Jean, A. Alexis, and R. Diehl. Galactic annihilation emission from nucleosynthesis positrons. *Astron.&Astroph.*, 543:A3, July 2012. doi: 10.1051/0004-6361/201118721.

I. F. Mirabel and L. F. Rodríguez. Sources of Relativistic Jets in the Galaxy. *ARA&A*, 37: 409–443, 1999. doi: 10.1146/annurev.astro.37.1.409.

I. F. Mirabel, L. F. Rodriguez, B. Cordier, J. Paul, and F. Lebrun. A double-sided radio jet from the compact Galactic Centre annihilator 1E140.7 - 2942. *Nature*, 358:215–217, July 1992. doi: 10.1038/358215a0.

U. Oberlack, K. Bennett, H. Bloemen, R. Diehl, C. Dupraz, W. Hermsen, J. Knoedlseder, D. Morris, V. Schoenfelder, A. Strong, and C. Winkler. The COMPTEL 1.809MeV all-sky image. *Astron.&Astroph. Suppl.*, 120:C311, Dec. 1996.

M. Pignatari, R. Gallino, M. Heil, M. Wiescher, F. Käppeler, F. Herwig, and S. Bisterzo. The Weak s-Process in Massive Stars and its Dependence on the Neutron Capture Cross Sections. *ApJ*, 710:1557–1577, Feb. 2010. doi: 10.1088/0004-637X/710/2/1557.

M. Pignatari, R. Hirschi, M. Wiescher, R. Gallino, M. Bennett, M. Beard, C. Fryer, F. Herwig, G. Rockefeller, and F. X. Timmes. The $^{12}C + ^{12}C$ Reaction and the Impact on Nucleosynthesis in Massive Stars. *ApJ*, 762:31, Jan. 2013a. doi: 10.1088/0004-637X/762/1/31.

M. Pignatari, M. Wiescher, F. X. Timmes, R. J. de Boer, F.-K. Thielemann, C. Fryer, A. Heger, F. Herwig, and R. Hirschi. Production of Carbon-rich Presolar Grains from Massive Stars. *ApJ*, 767:L22, Apr. 2013b. doi: 10.1088/2041-8205/767/2/L22.

S. Plüschke, R. Diehl, V. Schönfelder, H. Bloemen, W. Hermsen, K. Bennett, C. Winkler, M. McConnell, J. Ryan, U. Oberlack, and J. Knödlseder. The COMPTEL 1.809 MeV survey. In A. Gimenez, V. Reglero, and C. Winkler, editors, *Exploring the Gamma-Ray Universe*, volume 459 of *ESA Special Publication*, pages 55–58, Sept. 2001.

W. G. L. Pöppel, E. Bajaja, E. M. Arnal, and R. Morras. The interstellar medium surrounding the Scorpius-Centaurus association revisited. *Astron.&Astroph.*, 512:A83+, Mar. 2010. doi: 10.1051/0004-6361/200811290.

N. Prantzos and R. Diehl. Radioactive 26Al in the galaxy: observations versus theory. *Phys. Rep.*, 267:1–69, Mar. 1996. doi: 10.1016/0370-1573(95)00055-0.

N. Prantzos, C. Boehm, A. M. Bykov, R. Diehl, K. Ferrière, N. Guessoum, P. Jean, J. Knoedlseder, A. Marcowith, I. V. Moskalenko, A. Strong, and G. Weidenspointner. The 511 keV emission from positron annihilation in the Galaxy. *Reviews of Modern Physics*, 83: 1001–1056, July 2011. doi: 10.1103/RevModPhys.83.1001.

T. Preibisch, A. G. A. Brown, T. Bridges, E. Guenther, and H. Zinnecker. Exploring the Full Stellar Population of the Upper Scorpius OB Association. *Astron.Journ.*, 124:404–416, July 2002. doi: 10.1086/341174.

W. R. Purcell, L.-X. Cheng, D. D. Dixon, R. L. Kinzer, J. D. Kurfess, M. Leventhal, M. A. Saunders, J. G. Skibo, D. M. Smith, and J. Tueller. OSSE Mapping of Galactic 511 keV Positron Annihilation Line Emission. *ApJ*, 491:725–+, Dec. 1997. doi: 10.1086/304994.

R. Ramaty, M. Leventhal, K. W. Chan, and R. E. Lingenfelter. On the origin of variable 511 keV line emission from the Galactic center region. *ApJ*, 392:L63–L66, June 1992. doi: 10.1086/186426.

R. Ramaty, J. G. Skibo, and R. E. Lingenfelter. Galactic 0.511 MeV line emission. *ApJS*, 92: 393–399, June 1994. doi: 10.1086/191984.

R. A. Remillard and J. E. McClintock. X-Ray Properties of Black-Hole Binaries. *ARA&A*, 44: 49–92, Sept. 2006. doi: 10.1146/annurev.astro.44.051905.092532.

J. Rodriguez, M. Cadolle Bel, J. Alfonso-Garzón, T. Siegert, X.-L. Zhang, V. Grinberg, V. Savchenko, J. A. Tomsick, J. Chenevez, M. Clavel, S. Corbel, R. Diehl, A. Domingo, C. Gouiffès, J. Greiner, M. G. H. Krause, P. Laurent, A. Loh, S. Markoff, J. M. Mas-Hesse, J. C. A. Miller-Jones, D. M. Russell, and J. Wilms. Correlated optical, X-ray, and γ-ray flaring activity seen with INTEGRAL during the 2015 outburst of V404 Cygni. *Astron.&Astroph.*, 581:L9, Sept. 2015. doi: 10.1051/0004-6361/201527043.

F. K. Röpke, I. R. Seitenzahl, S. Benitez, M. Fink, R. Pakmor, M. Kromer, S. A. Sim, F. Ciaraldi-Schoolmann, and W. Hillebrandt. Modeling Type Ia supernova explosions. *Progress in Particle and Nuclear Physics*, 66:309–318, Apr. 2011. doi: 10.1016/j.ppnp.2011.01.026.

G. Rugel, T. Faestermann, K. Knie, G. Korschinek, M. Poutivtsev, D. Schumann, N. Kivel, I. Günther-Leopold, R. Weinreich, and M. Wohlmuther. New Measurement of the Fe60 Half-Life. *Physical Review Letters*, 103(7):072502–+, Aug. 2009. doi: 10.1103/PhysRevLett.103.072502.

V. Schoenfelder, K. Bennett, H. Bloemen, R. Diehl, W. Hermsen, G. Lichti, M. McConnell, J. Ryan, A. Strong, and C. Winkler. COMPTEL overview: Achievements and expectations. *Astron.&Astroph. Suppl.*, 120:C13+, Nov. 1996.

G. H. Share, R. L. Kinzer, J. D. Kurfess, D. J. Forrest, E. L. Chupp, and E. Rieger. Detection of galactic Al-26 gamma radiation by the SMM spectrometer. *ApJ*, 292:L61–L65, May 1985. doi: 10.1086/184473.

R. G. J. e. a. Siegert, T.; Diehl. Positron annihilation signatures associated with teh outburst of the microquasar v404 cygni. *Nature*, 531(7592):?, March 2016.

T. Siegert, R. Diehl, M. G. H. Krause, and J. Greiner. Revisiting INTEGRAL/SPI observations of ^{44}Ti from Cassiopeia A. *Astron.&Astroph.*, 579:A124, July 2015. doi: 10.1051/0004-6361/201525877.

T. Siegert, R. Diehl, G. Khachatryan, M. G. H. Krause, F. Guglielmetti, J. Greiner, A. W. Strong, and X. Zhang. Gamma-ray spectroscopy of positron annihilation in the Milky Way. *Astron.&Astroph.*, 586:A84, Feb. 2016. doi: 10.1051/0004-6361/201527510.

M. Signore and C. Dupraz. Massive stars as Galactic producers of Al-26. *Astron.&Astroph. Suppl.*, 97:141–144, Jan. 1993.

G. Skinner, P. Jean, J. Knoedlseder, P. von Ballmoos, M. Leising, P. Milne, and G. Weidenspointner. The 511 keV sky as seen by INTEGRAL/SPI, CGRO-OSSE and GRS/SMM combined. In *Proceedings of "An INTEGRAL view of the high-energy sky (the first 10 years)" - 9th INTEGRAL Workshop and celebration of the 10th anniversary of the launch (INTEGRAL 2012). 15-19 October 2012. Bibliotheque Nationale de France, Paris, France. Published online at ¡A href="http://pos.sissa.it/cgi-bin/reader/conf.cgi?confid=176"¿http://pos.sissa.it/cgi-bin/reader/conf.cgi?confid=176¡/A¿, id.112*, page 112, 2012.

L.-S. The and A. Burrows. Expectations for the Hard X-Ray Continuum and Gamma-Ray Line Fluxes from the Type Ia Supernova SN 2014J in M82. *ApJ*, 786:141, May 2014. doi: 10.1088/0004-637X/786/2/141.

F. X. Timmes, S. E. Woosley, D. H. Hartmann, R. D. Hoffman, T. A. Weaver, and F. Matteucci. 26Al and 60Fe from Supernova Explosions. *ApJ*, 449:204–+, Aug. 1995. doi: 10.1086/176046.

J. Trombka, C. Fichtel, J. Grindlay, and R. Hofstadter. Gamma-ray astrophysics - A new look at the universe. *Science*, 202:933–938, Dec. 1978. doi: 10.1126/science.202.4371.933.

J. I. Trombka and C. E. Fichtel. Gamma-ray astrophysics. *Phys. Rep.*, 97:173–218, 1983. doi: 10.1016/0370-1573(83)90116-3.

G. Vedrenne, J.-P. Roques, V. Schönfelder, P. Mandrou, G. G. Lichti, A. von Kienlin, B. Cordier, S. Schanne, J. Knödlseder, G. Skinner, P. Jean, F. Sanchez, P. Caraveo, B. Teegarden, P. von Ballmoos, L. Bouchet, P. Paul, J. Matteson, S. Boggs, C. Wunderer, P. Leleux, G. Weidenspointner, P. Durouchoux, R. Diehl, A. Strong, M. Cassé, M. A. Clair, and Y. André. SPI: The spectrometer aboard INTEGRAL. *Astron.&Astroph.*, 411:L63–L70, Nov. 2003. doi: 10.1051/0004-6361:20031482.

J. Vink. X- and γ-ray studies of Cas A: exposing core collapse to the core. *New Astron. Rev.*, 48:61–67, Feb. 2004. doi: 10.1016/j.newar.2003.11.008.

R. Voss, R. Diehl, D. H. Hartmann, M. Cerviño, J. S. Vink, G. Meynet, M. Limongi, and A. Chieffi. Using population synthesis of massive stars to study the interstellar medium near OB associations. *Astron.&Astroph.*, 504:531–542, Sept. 2009. doi: 10.1051/0004-6361/200912260.

R. Voss, R. Diehl, J. S. Vink, and D. H. Hartmann. Probing the evolving massive star population in Orion with kinematic and radioactive tracers. *Astron.&Astroph.*, 520:A51+, Sept. 2010. doi: 10.1051/0004-6361/201014408.

R. Voss, P. Martin, R. Diehl, J. S. Vink, D. H. Hartmann, and T. Preibisch. Energetic feedback and ^{26}Al from massive stars and their supernovae in the Carina region. *Astron.&Astroph.*, 539:A66, Mar. 2012. doi: 10.1051/0004-6361/201118209.

W. Wang, M. J. Harris, R. Diehl, H. Halloin, B. Cordier, A. W. Strong, K. Kretschmer, J. Knödlseder, P. Jean, G. G. Lichti, J. P. Roques, S. Schanne, A. von Kienlin, G. Weidenspointner, and C. Wunderer. SPI observations of the diffuse ^{60}Fe emission in the Galaxy. *Astron.&Astroph.*, 469:1005–1012, July 2007. doi: 10.1051/0004-6361:20066982.

G. Weidenspointner, G. Skinner, P. Jean, J. Knödlseder, P. von Ballmoos, G. Bignami, R. Diehl, A. W. Strong, B. Cordier, S. Schanne, and C. Winkler. An asymmetric distribution of positrons in the Galactic disk revealed by γ-rays. *Nature*, 451:159–162, Jan. 2008. doi: 10.1038/nature06490.

C. Weidner and P. Kroupa. The Variation of Integrated Star Initial Mass Functions among Galaxies. *ApJ*, 625:754–762, June 2005. doi: 10.1086/429867.

C. Weidner, P. Kroupa, and I. A. D. Bonnell. The relation between the most-massive star and its parental star cluster mass. *Mon.Not.Royal Astr. Soc.*, 401:275–293, Jan. 2010. doi: 10.1111/j.1365-2966.2009.15633.x.

C. Winkler, T. J.-L. Courvoisier, G. Di Cocco, N. Gehrels, A. Giménez, S. Grebenev, W. Hermsen, J. M. Mas-Hesse, F. Lebrun, N. Lund, G. G. C. Palumbo, J. Paul, J.-P. Roques, H. Schnopper, V. Schönfelder, R. Sunyaev, B. Teegarden, P. Ubertini, G. Vedrenne, and A. J. Dean. The INTEGRAL mission. *Astron.&Astroph.*, 411:L1–L6, Nov. 2003. doi: 10.1051/0004-6361:20031288.

Introduction to stellar evolution

Scilla Degl'Innocenti

Physics Department, Pisa University, Largo B. Pontecorvo n.3, I-56127 Pisa
INFN, Sezione di Pisa, Largo B. Pontecorvo n.3, I-56127 Pisa

E-mail: scilla.deglinnocenti@unipi.it

Abstract. This contribution is meant as a first brief introduction to stellar physics. First I shortly describe the main physical processes active in stellar structures then I summarize the most important features during the stellar life-cycle.

1. Introduction

The quantitative study of the physical structures called "stars" developed later with respects of other physical topics, as electromagnetism or thermodynamics. This because to understand the operation of stellar structures one must already be acquainted with fundamental physical mechanisms as photon-matter interactions or energy production through nuclear fusion. Moreover precise evaluations have been made possible only by the development of electronic calculators. At the same time telescopes more and more refined have been developed, also on board of satellites, to observe the whole electromagnetic spectrum without the limitations induced by the terrestrial atmosphere. Now the characteristics of stars of different mass, chemical composition and evolutionary phase are known with a quite good precision; the general scenario is well defined and in agreement with the most of observations of the Sun and of the stars in our Galaxy.

2. Stellar Observables

The main stellar observables are luminosity and color. The luminosity is the energy emitted per unit time, but it is not directly measurable for two main reasons. In general, the adopted detector (telescope) is not sensitive to the whole electromagnetic spectrum, but only to a given spectral range, which depends on its characteristics. Most important, due to the fact that a star isotropically emits its energy in all the directions, what is measured is a flux (energy per unit time and unit surface of the detector) and not a luminosity which, however, is the physical quantity directly linked to the stellar characteristics. In more detail, if L_* is the stellar luminosity, a detector at a distance d from the star measures the flux:

$$\Phi = \frac{L_*}{4\pi d^2} \tag{1}$$

The luminosity can be obtained only if a distance estimate is available; it's thus easy to understand why the problem of distance determination of astrophysical objects is of paramount importance. The color of a star, that is the spectral distribution of the emitted energy, can be directly connected with the temperature of the stellar surface (photosphere). In fact for

Journal of Physics: Conference Series **703** (2016) 012002 doi:10.1088/1742-6596/703/1/012002

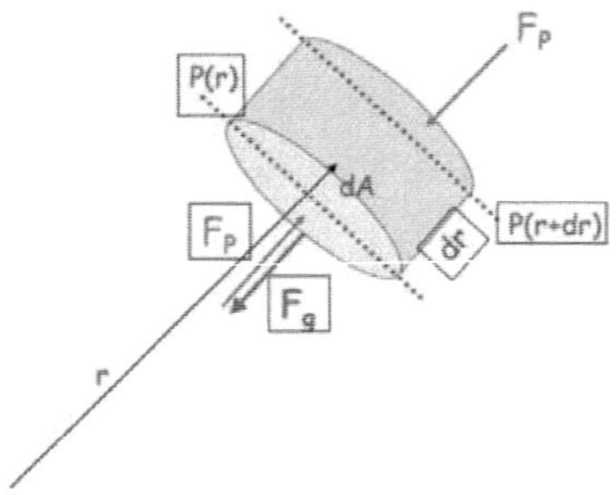

Figure 1. Balance between pressure force (F_P) and gravity force (F_g) on a given infinitesimal element at distance r from the center of the star with tickness dr and base area dA.

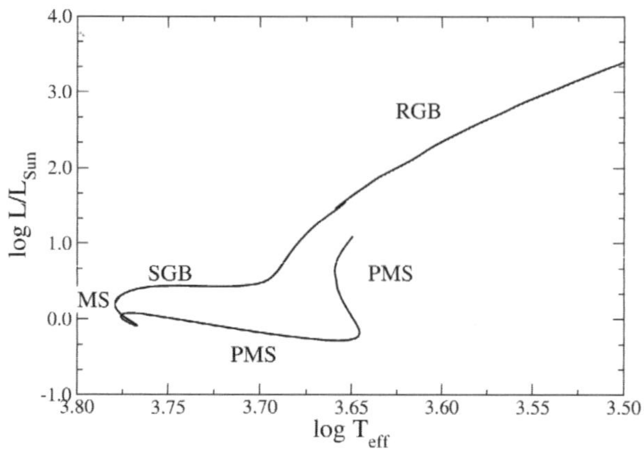

Figure 2. Evolution in the Hertzsprung-Russell diagram of a 1 $M_\odot$ star from the Pre-Main Sequence to the end of the Red Giant Branch phase.

almost the whole stellar structure photon-matter interactions are so efficient that the star can be assumed at thermodynamical equilibrium. In this case the gas particles follow a Maxwell-Boltzmann distribution while the photons a Planck one, with the same temperature for particles and photons. The temperature clearly varies with the position inside the star; in most cases the temperature monotonic decreases from the center to the stellar surface. Under these conditions the energy spectral distribution is the so called "black body distribution". In reality the emission of a star with a given surface temperature slightly differs from the one of a black body of the same temperature, due to emission and absorption phenomena by the atoms/molecules/grains of the stellar atmosphere. For a precise relation between the observed stellar color and the surface temperature one needs so solve the photons transport equation in the stellar atmosphere, taking into account all the possible absorption/emission phenomena. It's easy to understand that these calculations are very complex and they become more and more difficult as the photospheric temperature decreases because, due to the higher presence of molecules and grains, the possible photon-matter interactions increase more an more. These kind of calculations, which are an example of how much computing capacity is fundamental for precise stellar physics evaluations, are performed by few groups in the world and they are still affected by not negligible uncertainties mainly for stars with relatively low photospheric temperature. In absence of high rotation velocities and magnetic fields (which is supposed to be the situation of the most of the stars) the stellar equilibrium configuration is a sphere (as we will better discuss in the following); however, in general, it's not possible to directly measure the stellar radius. Stellar surface temperatures are generally expressed as "effective temperature" , that is the temperature which would have a star if it would exactly emit as a black body. Thus the relation among luminosity (L), radius (R) and effective temperature (T_{eff}) is:

$$L = 4\pi R^2 \sigma T_{eff}^4 \tag{2}$$

where σ is the StefanBoltzmann constant, $\sigma = 5.67 \cdot 10^{-5}$ dyne cm^{-2} K^{-4}. Usually, observed stars are shown in a plane which resumes the main stellar observables: effective temperature (or correspondingly the stellar color) on the abscissa and total luminosity (or correspondingly the luminosity in a given photometric band) on the ordinate, the so called Hertzsprung-Russell (Color-Magnitude) diagram.

3. Stellar structure and evolution equations

To "solve" a stellar structure means to theoretically calculate its physical quantities (temperature, pressure, density, luminosity production etc..) at each point inside the star. To do this one has to solve the so called "stellar equilibrium equations" which reflect the stellar characteristics describing its working principles.

3.1. Stellar gas characteristics

Stars are spheres of photons and plasma, that is ionized gas in which electrons are extracted from the atoms; the plasma is thus constituted by electrons and ions but it is globally neutral. The ionization state of the atoms which constitute the stellar plasma mainly depends on temperature and density (and clearly on the ionization energy of the various electrons of the different atomic species) and thus it varies inside the star. At each point in the stellar interior the ionization state of each specie can be calculated through the Saha equation [see e.g. 1, , cap.3]. The thermodynamical properties of the plasma can be calculated as a function of the chemical composition and of two physical quantities (e.g. pressure and temperature) through the equation of state (EOS) of a system of plasma and photons at thermodynamical equilibrium. For the Sun the law of perfect gas holds with a very good approximation, except for the very central regions in which few percents of the pressure are due to degenerate electrons. For different conditions of temperature and pressure the EOS can be more complicated; at relatively high density electrons became degenerate and under specific conditions coulombian effects among ions are no more negligible. Thus if one knows the chemical composition and two physical quantities (e.g. temperature and pressure) at a given point inside the star, one can obtain through the EOS all the thermodynamical quantities at that point. This can be synthetically written as:

$$\rho(r) = \rho[T(r), P(r), X(r), Y(r), Z(r)], \tag{3}$$

where X(r),Y(r) and Z(r) are, respectively, the fractional abundance in mass of hydrogen, helium and elements heavier than helium, which in astrophysics are called "metals", at a given radial coordinate. Clearly X(r)+Y(r)+Z(r)=1.

3.2. Hydrostatic equilibrium

Stellar lifetimes are orders of magnitude higher than the human ones, thus it's impossible to directly follow a stellar life-cycle. However now one can observe an enormous number of stars at different ages, obtaining in this way information about the characteristics of the stars at different phases of their life. Moreover the Sun has been observed in a scientific way from centuries. In addition the fact that the life is present on the Earth from some millions of years means that the solar characteristics did not change in a relevant way during this period. The evidence that most of the stars do not change their luminosity, effective temperature and thus radius for periods longer than centuries means that they are structures at mechanical equilibrium. The dynamical timescales (that is the time in which a structure out of dynamical equilibrium changes its characteristics) in stars are in fact of the orders of hours, [see e.g. 2, cap.1]. At each point inside a star the gas pressure counteracts the gravity force which would cause the stellar contraction. The gravity is a central force, thus the equilibrium configuration is a sphere. This simplifies very much the calculations: for spherical symmetry all the quantities can be expressed as a function of the radial coordinate. One can imagine a stellar structure composed of several spherical shells so thin that the physical quantities inside each shell can be assumed constant; by solving the stellar structure equations one can obtain the value of the physical quantities in each shell.

The mass, dM, inside a shell of radius dr is then:

$$dM(r) = 4\pi r^2 \rho(r) dr; \tag{4}$$

where ρ(r) is the corresponding density inside the shell. This is called "continuity equation".

If at each point of the star one equates the pressure force to the gravity force, as shown in Fig.1, one obtain the so called "hydrostatic equilibrium equation":

$$\frac{dP(r)}{dr} = -G\frac{M(r)}{r^2}\rho(r) \tag{5}$$

where P(r) is the pressure and M(r) is the mass inside a sphere of radius r; the above equation holds at each radial coordinate r.

The hydrostatic equilibrium equation also implies that the most internal regions, which must "support" the above layers, have an higher pressure. Thus, in general, the pressure is expected to monotonically decrease from the center to the surface. The two equations written until now, as a function of the radius, contains three unknowns: P(r), ρ(r) and M(r). P(r) and ρ(r) are linked through the equation of state, but, except some particular EOS, in this case, also the temperature (unknown) appears in the EOS. Thus the system is still not resolvable. This is expected, because the energy production and transport, which play an important role in stellar structures have still not taken into account.

3.3. Thermal equilibrium and energy production

Stars are very hot structures and thus they radiate, that is they emit photons, losing energy from the surface. The fact that the stellar effective temperature remains constant for longtime means that the stellar surfaces are at thermal equilibrium, that is the energy losses from the photosphere through photons emission are counterbalanced by an energy flux which reaches the photosphere from the underlying layer. If this would not happen the photosphere would change its temperature in few minutes. Actually the concept of "thermal equilibrium" is more general and it must be applied to the whole star; the energy which enters in each shell is equal to the one which exits from the same shell plus the energy which is possibly produced inside the shell itself.

The stellar photosphere is continuously supplied of energy by the more internal regions. For the majority of the stellar life this energy is produced by nuclear reactions; stellar interiors are so hot that fusion nuclear reactions of nuclei to form heavier elements can happen. The nuclear reactions most important for energy production are among charged nuclei. For these fusion reactions high temperatures (higher than about 10^6K) are needed because the nuclei must reach enough energy to pass the Coulom barrier through tunnel effect, reaching nuclear distances ($\approx 10^{-13}$ cm) in a way that nuclear reaction can happen, [see e.g. 3, cap.3]. The mechanism of nuclear fusion in stars has been described in detail in other talks at this school and thus I will not discuss it here. I only remind that the higher is the charge of reacting nuclei the higher is the threshold temperature for the ignition of the fusion. However the nuclear fusion is not the only possible energy source in stars; stars can produce energy through thermodynamical transformations of the gas linked to contractions and expansions of stellar regions, the so called "gravitational energy". Then stars lose energy also through neutrino production. Neutrinos have a mean free path larger than the stellar radius (except the case of very dense stellar cores of massive stars at the end of their life) and thus they escape from the stars without interacting with the stellar matter, that is without releasing their energy to the stellar structure. During nuclear fusions electron neutrinos are formed; in this case the only effect is that neutrinos take away a relatively small amount of the energy produced by the fusion, which is no more available to the structure. Most important for the stellar energy balance is when neutrinos are produced in regions in which nuclear fusions are not active. Different processes active for different ranges of density and temperature are responsible for this kind of neutrino production: neutrino photoproduction, neutrinos from couple production, plasma neutrinos etc...[see e.g. 4]. In these case the neutrino production constitutes for the stars a net energy loss and thus the regions in which these kind of neutrinos are produces cool down. Summarizing, in each region of the star energy production through nuclear fusions and/or thermodynamical transformations and/or neutrino energy losses can happen. Usually we define ε the energy produced/lost per unit time and unit mass due to the mechanisms described above, that is:

$$\varepsilon(r) = \varepsilon_{nuclear}(r) + \varepsilon_{gravitational}(r) - \varepsilon_{neutrinos}(r) \tag{6}$$

Each process of energy production depends on the physical quantities (density/pressure and temperature) and chemical composition which vary at each point of the star. In particular the various nuclear reaction rates can be written as analytical expressions, as a function of temperature, density and reagent abundances [see e.g. 5]. With these information it's possible to write the thermal equilibrium equation in each shell of the star:

$$\frac{dL(r)}{dr} = 4\pi r^2 \rho(r)\varepsilon(r); \tag{7}$$

where L(r) is the luminosity (energy per unit time) which enters in a shell of radius r. Another equation is added to the system but another unknown, i.e L(r), is also added.

3.4. Energy transport

The energy produced in the internal regions must be transferred to the whole star. The only energy transport mechanism which is always active in stars is the radiative transport, that is the totality of processes of photon-matter interaction. The main processes in the stellar interiors are: photon-electron scattering (which, due to the relatively low energies, is mainly a Thompson scattering), photoionization, inverse bremsstrahlung, photon absorption (and further isotropic emission) of photons by atoms ,[see e.g. 1, cap.16]. While for the external parts of the stars, much less dense and with relatively low temperatures, the main opacity sources are scattering and photoionization on H^- ions, absorption on molecules and grains, scattering on molecules and molecules photodissociation. The total amount of photon-matter interaction processes, which remove energy from the outgoing flux, is called “opacity”, k, and it is measured as a cross section (of the various processes) per unit mass, that is in cm^2/g. In this sense the induced emission process constitutes a negative contribution to opacity. Clearly, each opacity process depends on the physical quantities (e.g. temperature and density) and on the chemical composition, in a way that not only the total opacity varies with radius but also the photon matter interaction process which dominates the opacity is dependent on the position inside the star. The opacity also depends on photon distribution. By assuming a Planck distribution (which is a valid guess for the whole star, but the atmosphere) one can mediate the opacity over the photon distribution obtaining the so called “Rosseland mean opacity” , $\bar{k}$,[see e.g. 2, cap.4]. It's easy to understand that opacity calculations are very difficult because one must evaluate the efficiency of the described processes for each element of the stellar mixture taking into account all the possible ionization states. Opacity calculations must thus rely on a equation of state too. Few groups in the world are able to perform these calculations; the results are made available to the astrophysical community through tables as a function of temperature, density and chemical composition. By imposing that photons follow the Planck distribution and assuming that the photon-matter interaction is the only active energy transport mechanism, it's easy to find [see e.g. 6, cap.3], at each point inside the star, a relation between the temperature gradient and energy flux to be transported, the so called “energy radiative transport equation”:

$$\frac{dT(r)}{dr} = -\frac{3}{4}\frac{\bar{k}(r)\rho(r)}{ac}\frac{1}{T(r)^3}\Phi(r); \tag{8}$$

where the photon flux Φ at each point of the star is:

$$\Phi(r) = \frac{L(r)}{4\pi r^2}, \tag{9}$$

c is the light speed and a is the radiation pressure constant (a=$7.56 \cdot 10^{-15}$ erg cm^{-3} K^{-4}). Thus one sees that the temperature gradient needed to transport a given flux entirely through photon matter interactions (radiative gradient) is proportional to the flux itself and to the opacity, that is to the "resistance" of the matter to the photon flux. For completeness one has to mention that other energy transport mechanisms are possible. If the temperature gradient required to a stellar structure to transport energy through photon-matter interactions would be too much high, energy transport through convection becomes efficient too, even if radiative transport is still present. Briefly, the convective energy transport is due to mass elements which move from hotter to colder regions releasing their heat similarly to what happens for the water boiling in a pot. In this case the total energy flux is transported through both the radiative and convective mechanisms and the temperature gradient expression is different from the one written in eq.8. A precise treatment of convection is out of the purposes of this talk but it's useful to remind that convection is active in the regions where the radiative gradient of eq.8, which is proportional to the energy flux and to the opacity, is enough high. More precisely, for the convection activation, the radiative gradient must be higher than the adiabatic gradient, a thermodynamical quantity, obtained from the EOS. It's also worth noticing that in very dense stellar regions energy transport due to degenerate electron conduction [see e.g. 2, cap.4] dominates; in this case too eq.8 is no more valid.

If one reminds that pressure is, in general, monotonically decreasing from the center to the surface, taking into account the radiative transport equation and the thermodynamical characteristics of the photon and gas mixture (e.g. for simplicity one can image a perfect gas law) generally speaking one can understand that temperature and density too decreases from the center to the stellar surface.

We have thus written in total four equations as a function of the radius for four unknowns (pressure, temperature, mass, luminosity), whereas the density is related to pressure and temperature through the EOS; thus the system is resolvable, once ε(r) and $\bar{k}$(r) are evaluated. These differential equations can be numerically solved [see e.g. 6, cap.3] finding, once fixed the mass and the stellar original chemical composition, the physical quantities in each point of the star.

As it will be better discuss in the following, the temporal evolution of the stars is generally driven by chemical composition variations, mainly due to nuclear reactions. Once solved the equations set at a given time, t_0, taking into account the variations of the chemical composition after a temporal step Δt, one can integrate again the equations with the new composition at the time $t_0 + \Delta t$ obtaining the stellar characteristics at that time. A similar procedure is applied in the phases in which the nuclear reactions are not active when the stellar changes are mainly driven by the contraction/expansion of the structure. These procedures are iterated for several temporal steps to follow the stellar evolution.

4. Stellar evolution

In the following I will describe the main stellar features during the different evolutionary phases, as obtained by solving the stellar structure and evolution equations; some of the main characteristics can also be roughly obtained by general evaluations discussed above.

4.1. Pre-Main Sequence and Main Sequence stars

Stars born inside a molecular cloud of gas and interstellar dust; the gas is mainly composed by H and He with traces of other elements. These molecular clouds, neglecting, as a first

Journal of Physics: Conference Series **703** (2016) 012002 doi:10.1088/1742-6596/703/1/012002

approximation, possible effects due to rotation and magnetic fields, are in equilibrium between the gravitational force and the pressure force. However, due to stochastic motions, the colder and more dense regions of the cloud, can start to contract. As soon as the collapse proceeds the region fragments more and more; until each part in which the original region of the cloud has been divided will form a star. The way in which the gas of the original cloud falls onto the star until the achievement of the stellar final mass is still an open problem, because hydrodynamical protostellar models are still largely debated and not yet settled. The duration of the protostellar phase too is not known with precision and it varies with the stellar mass, but it is supposed to range from few 10^5 to about 10^6 years, [see e.g. 7, for details].

Anyhow, at the end of the protostellar phase the radiation pressure of the new born star sweep away the residual gas of the original cloud. The star is now in the "Pre-Main Sequence phase (PMS)". Moreover during the first phases of their life stars are supposed to be fully convective; thus they are completely mixed and possible chemical inhomogeneities of the original cloud are canceled. During the protostellar phase the star contracted and warmed up until the achievement of the hydrostatic equilibrium: the gas force exactly counterbalances the gravity force. The stellar gas pressure is mainly due to its high temperature. However the star is warm and radiates, loosing energy from the surface; thus, after a while, the star cools down and its pressure decreases becoming insufficient to completely sustain the star against the gravity. The star slightly contracts increasing its temperature and reaching again the hydrostatic equilibrium until the radiation losses from the surface cools down again the star. The PMS phase is characterized by these consecutive stellar contractions when the star passes from one "quasi-equilibrium" state to another one, while its internal temperature increases. The timescale of these processes is the one at which the stars cools down, that is the timescale in which a photon produced in the stellar interior can reach the surface. This timescale that can be seen also as the "relaxation time" for departure of a star from the thermal equilibrium (the so called "Kelvin-Helmholtz" time) varies with the stellar mass but it is of the order of 10^6-10^7 years.

While the star contracts its density increases too and the radius significantly decreases, then also the luminosity strongly decreases as shown in Fig.2. During the PMS evolution a star reaches first the temperature for deuterium and light elements (Li, Be, B) burning, then it reaches the temperature for hydrogen burning into helium in its core, which is the warmest region of the star. The hydrogen fusion into helium is the first nuclear reaction able to produce an energy which can replace the radiative losses from the surface. Hydrogen is the first abundant element to be burned because the coulombian barrier for the fusion of two protons is the lowest one and H is the most abundant element in the Universe. When the hydrogen fusion is fully active the star is in equilibrium, with the nuclear reactions which supply to the star the energy lost by photon emission from the photosphere. Sometimes one hears on TV broadcasts that the stellar luminosity depends on the efficiency of the nuclear reactions, but actually the exactly opposite is true. Stars are luminous because are warm and, as all the warm bodies, they irradiate. As a first approximation, the stellar temperature depends from the fact that the stellar gas must be warm to have enough pressure to counteract the gravity force; then the surface temperature also depends on the temperature gradient inside the star (see eq.8). Thus, roughly speaking, the higher is the mass of the star, the higher is its effective temperature and its luminosity. When, at the end of the PMS phase, hydrogen fusion ignites, it adjust its efficiency until the energy losses from the surface, proportional to the effective temperature and the stellar radius, are completely counterbalanced. This because in stars a feedback mechanism is efficient: if the produced energy is lower than radiative losses the star still cools down and contract, even if with a longer timescale. The result is the increase of the temperature of the central regions together with the burning efficiency, which is very sensitive to the temperature. On the contrary if the produced energy is higher than the one lost from the surface, the central regions warm up and expand; this leads to a temperature decrease and thus to a reduction of the burning efficiency. At the end of the process the energy produced by nuclear reactions is adjusted to exactly counterbalance the surface energy losses. Thus, in this phase, the stellar luminosity is mainly due to the stellar mass (with a smaller dependence on the chemical composition) and the nuclear reaction efficiency is tuned to exactly produce the energy needed to the star to irradiate. When nuclear reactions are so efficient to counterbalance the energy losses from the surface, the star leaves the PMS phase starting its Main Sequence (MS) evolution. The name "Main Sequence" is due to the fact that stars in this phase occupy in the Hertzsprung-Russell diagram a mono-parametric sequence (neglecting the secondary effect of the chemical composition) whose parameter is the stellar mass. It's worth noticing that stars with masses lower than about 0.08 $M_{\odot}$ do not succeed to reach the central temperatures needed to hydrogen ignition. They remain in the Pre-Main Sequence phase continuing to contract while the density increases until the star is sustained by the pressure of the degenerate electrons; these kind of stars are called "brown dwarfs". The star is now in equilibrium thanks to the nuclear energy, but the reaction efficiency decreases according to the reduction of the fuel (hydrogen) abundance. Thus, after a while, the nuclear energy is no more enough to counterbalance the radiative losses and the star contracts again, with the timescale of the fuel abundance reduction, that is of the nuclear burning rate. Thus the temperature of the central regions slightly increases until the energetic equilibrium is reached again, with a slightly different configuration of the stellar structure; that is the star evolves with nuclear timescale.

Nuclear timescales strongly vary with the stellar mass; the higher is the mass, the higher is the luminosity and thus the higher is the rate of the nuclear reactions which counterbalance the energy losses due to the luminosity. Moreover the higher is the fusion rate, the lower is the time in which the fuel is exhausted. To be more precise for the most of the stars the luminosity increases with the mass to a power of $2 \div 4$, while the H abundance linearly increases with the mass. As an example, the MS time for the Sun is about 10 Gyr while for a 15 $M_{\odot}$ star of the same chemical composition is about 10^7years.

4.2. Advanced evolutionary phases

The history of a star after the central H exhaustion can be roughly seen as a succession of central burnings of elements with charges more and more high and central regions contractions to possibly reach the temperatures needed for the various burning ignitions. In the phases in which the stellar core contracts nuclear burnings in a shell out of the central regions are often present. One has to remind that the higher is the nuclear charge of the reacting nuclei the higher is the coulombian barrier and thus the higher is the temperature needed for the fusion ignition. But the higher is the stellar mass the higher is its central temperature. The logical consequence is that only the most massive stars ($M \approx 10\ M_{\odot}$) can reach the temperatures needed to ignite all the exoenergetic reactions up to the production of the elements of the iron peak. These stars can thus produce all the elements of the periodic table, through fusions of charged particles and neutrons and protons captures on nuclei. The death of these stars is through an explosion in which the most of the stellar matter is ejected in the space (type II supernova) and thus it will constitute the cradle for the birth of a following stellar generation. Stars with lower masses have different final fates in dependence of their mass.

After the central H burning the stars experience a phase in which H is burned in a shell around a central inert He core. In this phase the stellar envelope expands and cools down and the star moves toward the red in the HR diagram (Sub Giant Branch, SGB, phase), see Fig.2. During the MS evolution the burning core slightly contracted and its temperature increased, so that during the Sub Giant Branch H burning mainly occurs through the CN-NO bi-cycle, much more sensitive to the temperature than the concurrent proton-proton chain [see e.g. 3, cap.6]. When CNO hydrogen burning reaches the equilibrium of secondary elements in the cycle, the stars enter in the Red Gian Branch, RGB, phase (see Fig.2). Stars now have an extended and low density envelope; their luminosity and lifetime are mainly governed by the dense He core and the surrounding H burning shell, while the temperature gradient in the envelope determines the stellar effective temperature. In particular the RGB luminosity increases with the He core mass which grows due to the burning in the H shell which surrounds the He core. The RGB phase ends when the central regions of the star reach the temperature for helium burning into C and O ($\approx 10^8$ K), thanks to the He core contraction due to its mass growth. Stars with masses lower than about 2.3 $M_\odot$ (the so called "RGB transition" mass), whose interiors did not reach very high temperatures, remain longtime in the RGB phase because the reaching of the He burning temperature is more difficult. These stars reaches very high luminosities along the Red Giant Branch and thus the mass losses by stellar winds, driven by the radiation pressure, increase too. At the end stars with masses greater than about 0.5 $M_\odot$ succeed to ignite central He burning. The evolutionary theory predict that lower masses will remain in the RGB phase loosing more and more mass until the only remain will be the dense, degenerate He core, surrounded by a very thin H shell with all the nuclear burnings switched off. A similar structure, called He white dwarf (He WD), is maintained in equilibrium by the pressure of degenerate electrons which counteracts the gravity and it will cools down until it will reach the cosmic background temperature. However the Main Sequence lifetime of stars with $M < 0.5\ M_\odot$ is higher than the present age of the Universe, thus no one of these stars still reached the white dwarf phase. However He white dwarfs are observed; their progenitors are stars of mass higher that $0.5 M_\odot$, which during the RGB phase loose so much mass that the H shell switches off and the stars leave the RGB phase ending their life as He WDs. Stars which ignite He in the center but did not explode as supernovae end their life as white dwarfs with a C/O core or with a Ne core if they succeed to ignite carbon too.

Acknowledgments
I would like to acknowledge the collegues with which I am working in Pisa from many years in the stellar physics field: M. Dell'Omodarme, P.G. Prada Moroni, E. Tognelli and G. Valle.

References

[1] Cox J P 1968 *Principles of stellar structure - Vol.1: Physical principles; Vol.2: Applications to stars* (New York: Gordon and Breach)

[2] Hansen C J, Kawaler S D and Trimble V 2004 *Stellar interiors : physical principles, structure, and evolution* (New York: Springer-Verlag)

[3] Rolfs C E and Rodney W S 1988 *Cauldrons in the cosmos: Nuclear astrophysics* (Chicago, IL: University of Chicago Press)

[4] Raffelt G G 1996 *Stars as laboratories for fundamental physics : the astrophysics of neutrinos, axions, and other weakly interacting particles* (Chicago: University of Chicago Press)

[5] Fowler W A, Caughlan G R and Zimmerman B A 1967 *Annual Review of Astronomy and Astrophysics* **5** 525

[6] Salaris M and Cassisi S 2006 *Evolution of Stars and Stellar Populations* (Wiley-VCH)

[7] Stahler S W and Palla F 2005 *The Formation of Stars* (Wiley-VCH)

Neutron sources and neutron-capture paths in asymptotic giant branch stars

Maria Lugaro

Konkoly Observatory, Research Centre for Astronomy and Earth Sciences, Hungarian Academy of Sciences, H-1121 Budapest, and Monash Centre for Astrophysics, Monash University, VIC3800, Australia

E-mail: `maria.lugaro@csfk.mta.hu`

Abstract.
Roughly half of the abundances of the elements heavier than iron in the cosmos are produced by *slow* neutron captures (the *s* process) in hydrostatic conditions when the neutron density is below roughly 10^{13} n/cm^{-3}. While it is observationally well confirmed that asymptotic giant branch (AGB) stars are the main site of the *s* process, we are still facing many problems in the theoretical models and nuclear inputs. Major current issues are the effect of stellar rotation and magnetic fields and the determination of the rate of the neutron source reactions. I will present these problems and discuss the observational constraints that can help us to solve them, including spectroscopically derived abundances, meteoritic stardust, and stellar seismology. Further, I will present evidence that the *s* process is not the only neutron-capture process to occur in AGB stars: an *intermediate* process is also required to explain recent observations of post-AGB stars.

1. Introduction

Atomic nuclei in the Universe cover a huge variety of chemical properties, masses, abundances, and origins: from the very light H and He, which were made during the Big Bang, to Be, B, and Li, which are mainly produced by spallation reactions in the interstellar medium, to the most abundant "metals" C, N, and O, which make up most of the 1.4% of the matter in the Sun that is not in the form of H and He ("metallicity"). Beyond them, there are "intermediate-mass" elements, such as Mg, Na, Si, and Ca mainly produced in core-collapse supernovae. The "Fe-peak" includes elements from Mn to Ni, which have relatively high abundances and are produced by nuclear statistical equilibrium in thermonuclear supernovae. Once we get to Fe, however, we have only covered about 1/4 of the total number of the elements in the cosmos: the vast majority of the atomic nuclei lie between Fe and the actinides, e.g., U. In the Sun, the abundances of the elements heavier than Fe are roughly five to six orders of magnitudes lower than those of Fe (Figure 1). However, closer to us, in the Earth's crust their abundances are only two to three order of magnitude lower than that of Fe. Most importantly, among them are found major industrial metals (Mo, Sn, Pb), precious metals (Ag, Au, Pt), and the rare earth elements (Ce, La, Nd, Sm, etc), i.e., the seeds of technology elements that make possible the high-tech world we live in today. In this paper, I will review our current knowledge of how a fraction of the cosmic abundances of the elements heavier than Fe (including some of those that are used to build for example a smartphone) is produced in giant stars.

Beyond Fe, the Coulomb barrier generated by the protons in the atomic nucleus is too large to allow captures of charged particles, and the only effective way to produce nuclei heavier the Fe are neutron captures, since neutrons do not have a charge. Traditionally, neutron-capture processes are split into two extreme cases [1]: if the neutron density N_n is very low, of the order of 10^8 n/cm^3, the time that it takes for an unstable nucleus to capture a neutron is *slower* than the time it takes to decay. This is the case of the *slow* neutron-capture process (the *s* process), whose path effectively proceeds along the valley of β-stability. At the opposite extreme, when N_n is very high, $> 10^{20}$ n/cm^3, the time that it takes for an unstable nucleus to capture a neutron is more *rapid* than the time it takes to decay. This is the case of the *rapid* neutron-capture process (the *r* process), whose path effectively proceeds on the neutron-rich side far away from the valley of β-stability. At the end of the neutron flux, the *r* process has produced only radioactive nuclei, which β decay towards their first stable isobar. The *r* process to occur needs extreme conditions, which are found in neutron star mergers and supernovae environments (see paper by K. L. Kratz, these proceedings). Instead, the *s* process can happen during the hydrostatic evolution of massive stars, as well as of low-mass stars, during their asymptotic giant branch (AGB) phase. This final phase of the life of low-mass stars is presented in details in the paper by M. El Eid (these proceedings), so here I will focus mostly on the *s* process occurring within it. First, I will summarize the nuclear physics features of the *s* process, which are valid independently of the stellar site where it occurs.

1.1. The three s-process Golden Rules

To interpret models and observations of *s*-process patterns, to a first approximation, it is possible to rely on three simple rules.

(i) **Rule 1: Neutron magic nuclei on the *s*-process path have low neutron-capture cross sections, σ, relatively to the other nuclei. They act as bottlenecks, and can accumulate.** Because the *s*-process path runs along the valley of β-stability, these neutron magic nuclei are all stable. The first neutron magic number on the *s*-process path, which usually starts with ^{56}Fe as the seed, is 50 and the first magic nucleus is ^{88}Sr, which has σ = 6.13 mbarn (at 30 keV, see the kadonis.org database [2]). For comparison, ^{87}Sr has σ = 97 mbarn. Moving forward, the second neutron magic number is 82 at ^{138}Ba (σ = 4 mbarn) and the third is 126 at ^{208}Pb, which also has a magic number of protons (82, and σ = 0.36 mbarn). Because the neutron-capture cross section σ represents the probability that the nucleus will capture a neutron, if this probability is low, as for the neutron magic nuclei, then these nuclei are produced, but not destroyed as easily. A large number of "events", in the form of neutrons hitting them, are required to "convince" them to capture one. After a given number of "neutron hitting events", the magic nuclei will start capturing neutrons and the *s*-process flux will pass through and beyond them to produce heavier nuclei. This will happen sequentially, as the bootleneck at ^{87}Sr needs to be bypassed first, and the bootleneck at ^{138}Ba second. The bottleneck at ^{208}Pb is effectively never bypassed because all nuclei beyond it decay back so that the *s* process terminates here. More neutrons simply results in more ^{208}Pb.
The key quantity that determines if the magic bottlenecks are bypassed, and consequently what is the final *s*-process distribution, is clearly the number of "events", i.e., the number of times that the magic nucleus is hit by a neutron. This is proportional to a quantity called "neutron exposure", or τ, which is the total neutron flux integrated over time, i.e., $\int N_n v_{thermal}\,dt$, where $v_{thermal}$ is the velocity of the neutrons, equal to the thermal velocity. Unsurprinsingly, the distribution of the abundances in the Solar System shown in Figure 1 presents three *s*-process peak corresponding to the stable nuclei with neutron magic numbers at Sr, Ba, and Pb. The three *r*-process peaks at Se, Xe, and Pt corresponds instead to unstable nuclei with neutron magic numbers, since the *r*-process path runs far away from

vallye of β-stability (the element Sn also appears as a relative peak because it has a magic number of protons and some of its isotopes have relatively low σ values).

(ii) **Rule 2: In-between neutron magic nuclei the abundances are in equilibrium** because the more they are produced, the more they are destroyed (the reaction rate is proportional to the abundance itself) until production and destruction balance each other. (This is also true for neutron magic nuclei for high neutron exposures, i.e., once their bottlenecks are completely bypassed.) It can be derived that steady-state equilibrium on the s-process path is given by $N_s(A)\ \sigma(A) \sim$ constant, where $N_s(A)$ is the s-process abundance of a nucleus of mass A and $\sigma(A)$ its neutron-capture cross section. This means that it is very easy to calculate relative abundance as ratios of neutron-capture cross sections, e.g., $N_s(A)/N_s(A+1) \sim \sigma(A+1)/\sigma(A)$. To a first approximation, these are independent of the neutron flux, as well as of the thermodynamic conditions, however, one should keep in mind that some neutron-capture reaction rates $\sigma\, v_{thermal}$ may have a dependence on the temperature, in which case σ is not proportional to $v_{thermal}$, and the abundance ratios change with the temperature.

(iii) **Rule 3: "Branching points" can open, where the *s*-process path has the option to move slightly off the valley of β-stability, not much further than a few mass numbers. As a rule of thumb this may happen when the unstable nucleus has a half life larger than roughly a day.** While the classical definition of the s process involves very low neutron densities, $\sim 10^8$ n/cm^3, reality is that neutron captures still more or less follows the valley of β-stability for neutron densities up to $\sim 10^{13}$ n/cm^3. The difference is that as the neutron density progressively increases the probability of unstable nuclei to capture a neutron instead of decaying also increases, depending on their decay rate and σ and of course the neutron density. The decay rate can also depend on the temperature, in which case the branching point act as thermometers for the s process.

To apply the three Golden Rules, especially Rule 1 and Rule 3, one needs to understand the difference between the "neutron exposure" and the "neutron density". The neutron exposure controls Rule 1 while the neutron density controls Rule 3, so it is crucial to be able to distinguish the two. A simple explanatory diagram is shown in Figure 2. The neutron exposure is proportional to the total number of neutrons integrated over time, which is represented by the area below the neutron density profile. The neutron exposure is controlled by both the neutron density and the timescale of the neutron release. It possible to have cases where the neutron density is low, but the timescale is long and the neutron exposure is large (case *a* in the figure) and cases where the neutron density is high but the timescale is short, and the neutron exposure is small (case *b* in the figure). In fact, these two different cases corresponds for example to the activation of the two neutron sources in AGB the ^{13}C(α,n)^{16}O and the ^{22}Ne(α,n)^{25}Mg reactions discussed below.

2. The *s* process in AGB stars

Observational evidence of the occurrence of the s process in AGB stars has been accumulated starting from the 1950s [3, 4, 5, 6] and includes the observation of Tc, a radioactive elements that would not be present in the envelopes of these star if it was not produced *in situ* in their deep layers and mixed up to the surface. The structure and evolution of AGB is covered in the paper by M. El Eid (these proceedings), here, I briefly remind the essential points. An AGB stars is made essentially of two components: a large convective envelope that extends to hundreds of solar radii and includes most of the stellar mass, and a very compact inner region, which includes 1/3 to 1/6 of the total mass concentrated on a radius just a few times that of the Earth. The inner region is separated in different layers: most of the mass is located in an electron-degenerate core made of C and O from the previous H and He burning, on top of the core are the He- and

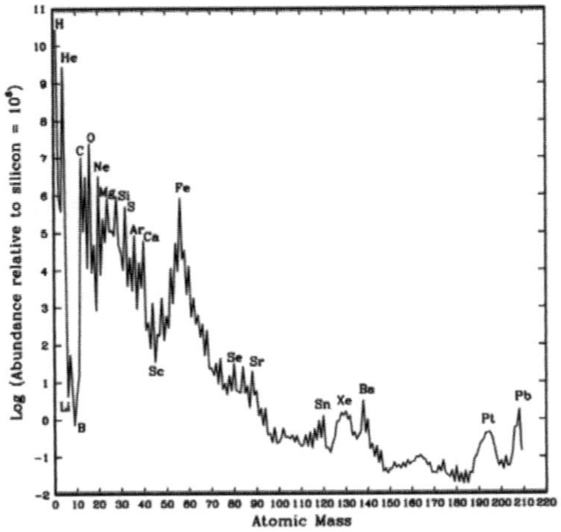

Figure 1. Solar abundances of atomic nuclei as function of their mass, normalised to Si=10^6. The elements corresponding to the different peaks are highlighted, in particular Sr, Ba, and Pb represent the three *s*-process peaks corresponding to the neutron magic numbers 50, 82, and 126.

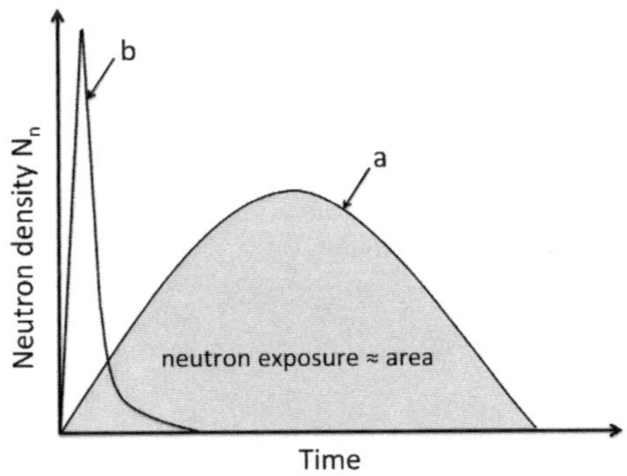

Figure 2. Schematic example of two possible temporal evolution of the neutron density during the *s* process, to illustrate the difference between neutron exposure and neutron density. The neutron exposure is proportional to the total number of neutrons released in time, i.e., to the area below the neutron density profile. The neutron exposure is much larger in case *a* than in case *b*, conversely, the neutron density reaches much higher values in case *b* then in case *a*.

H-burning shells separated by a region rich in He and called "intershell". This is the location of interesting nucleosynthesis, including the *s* process. The schematic temporal evolution of the internal structure of an AGB star is illustrated in Figure 3. The H-burning shell is activated most of the time, while the He-burning shell becomes active episodically and drives convection over the whole intershell ("convective pulse"). After a pulse is extinguished a mixing episode ("dredge-up") can occur between the envelope and the intershell carrying to the surface material synthetised by nuclear reactions in the intershell, e.g., C made in the pulse by partial He burning and *s*-process nuclei. This cycle can occur many times over the AGB lifetime, depending on the initial stellar mass and how quickly the strong stellar winds carry away the envelope material.

For the *s* process free neutrons are needed and the ^{13}C(α,n)^{16}O reaction is the main source in AGB stars because it activates already at temperatures around 90 MK. Partial mixing of protons is assumed in the models to occur in a thin region of the intershell at the deepest extent of each dredge-up episode where a sharp discontinutity arises between the intershell and the convective envelope. The protons react with ^{12}C and, under the basic assumption that the number of mixed protons declines with the mass depth, a "pocket" is produced, rich in ^{13}C in the bottom layers - where the mixed number of protons is lower - and rich in ^{14}N in the top layers - where the mixed number of protons is higher and CNO cycle equilibrium is achieved. The ^{13}C(α,n)^{16}O is activated usually before the onset of the next convective pulse, hence in radiative conditions creating a thin layer extremely *s*-process rich (Figure 3) [7, 8, 9, 10]. The initial stellar mass at which this sequence of events is believed to occur is between roughly 1.5 and 4 $M_\odot$. Below this mass range no dredge-up episodes are found, which are the premise for the assumed mixing of protons. Above this mass range, two effects play a role to remove the ^{13}C-rich pocket: (i) the mass of the intershell becomes progressively smaller, e.g., for a 2 $M_\odot$ star it is of the order of 0.02 $M_\odot$, while for a 5 $M_\odot$ star it is of the order of 0.001 $M_\odot$, so the

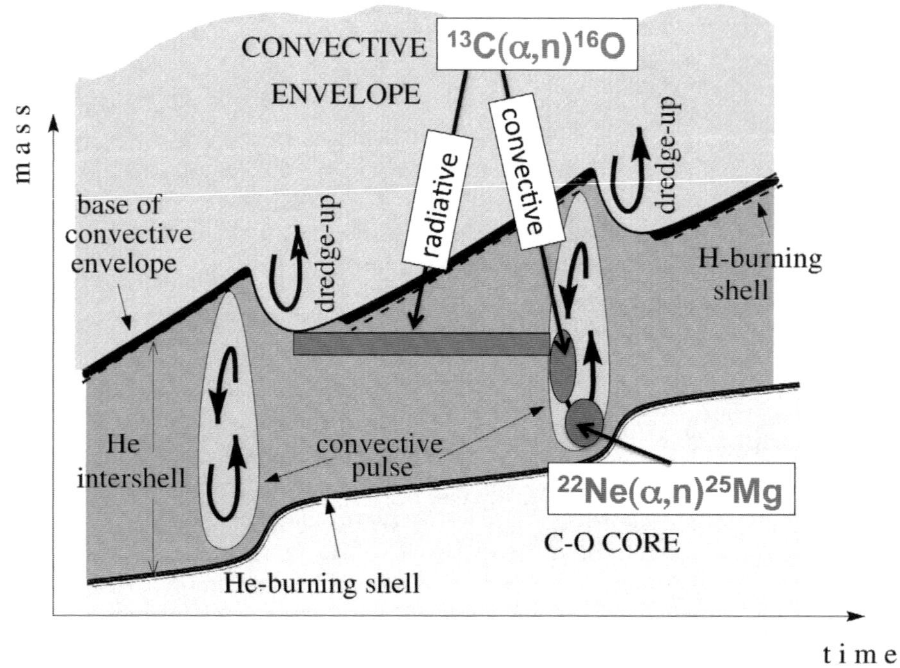

Figure 3. Schematic illustration of the evolution of the internal structure of an AGB star with indicated the locations in time and space where neutrons are released for the *s* process.

mass affected by the mixing of protons also shrinks [10]; (ii) the base of the envelope is so hot that the protons burn while they are mixed, and it is not possible to build a proton profile with a low abundance of protons in the layer deepest in mass leads to the production of ^{13}C [11]. This theoretical scenario is also confirmed by observations [12].

Currently, one of the most debated effects on the neutrons released in the ^{13}C pocket is that of stellar rotation. Initially, a star like the Sun can be assumed to rotate more or less like a rigid body, however, when it becomes a giant the envelope expands and slows down (like the ice skater opening his arms) and the core contracts and speeds up (like the ice skater closing his arms). The pocket is located at the border between these two different components, i.e., right on a discontinuity of the rotational velocity. This discontinuity generates mixing inside the otherwise radiative pocket, which carry ^{14}N from the top layers into the ^{13}C-rich bottom layers. ^{14}N is the main "neutron poison" for the *s* process, i.e., it steals neutrons from ^{56}Fe and the heavier nuclei because its ^{14}N(n,p)^{14}C cross section is relatively high, 1.83 mbarn - by comparison the neutron-capture cross sections of the C, N, O isotopes are of the order of 10^{-2} mbarn. In this situation the amount of free neutrons (and hence the neutron exposure) inside the ^{13}C pocket can decrease dramatically and the overall *s*-process distribution shifts towards lighter elements (Golden Rule 1) [13, 14, 15]. On the other hand, physical mechanisms exist that can couple the core to the envelope, make it rotate slower, and decrease the rotational velocity discontinuity and the effect of rotation on the *s* process. These are, for example, magnetic fields (both fossil and generated by rotation itself [16]) and gravity waves. We do not have yet an accurate description of their impact on the evolution of angular momentum in stars, but we do have observations of the rotational velocities of giant stars and of white dwarfs from stellar seismology [17, 18]. From trying to match these observations Cantiello et al. [19] derived that the AGB evolution should occur at nearly constant core angular momentum. It will be of much interest to investigate what is the effect of these results on the *s* process.

In summary, rotation can alter the neutron flux by generating mixing inside ^{13}C pockets, which would otherwise behave radiatively, and potentially producing a lower neutron exposure. Another situation where ^{13}C burns while mixed with ^{14}N occurs in stars of low masses and relatively high metallicity. In these stars the temperature in the ^{13}C pocket may not reach the 90 MK required to burn ^{13}C before the onset of the next convective pulse and the ^{13}C nuclei burn completely or partially as they are ingested in the convective thermal pulse (Figure 3) [10, 20]. With respect to the case where all the ^{13}C nuclei burn radiatively, the result is a lower neutron exposure - similarly to the rotation case, because the ^{13}C nuclei are mixed with ^{14}N nuclei, which in the case of the ingestion are also present in the H-burning ashes - but a higher neutron density - because the temperature is higher in the pulse and the burning timescale is a few tens of years rather than a few tens of thousands of years. The neutron profile looks more like case *b* in Figure 2, while in the radiative case it looks more like case *a*. Both the overall distribution (Golden Rule 1) and the effect of branching points (Golden Rule 3) are affected, with the overall distribution weighted towards the first *s* process peak and the branching points being more open.

2.1. Uncertainties in the $^{13}C(\alpha,n)^{16}O$ reaction and the impact on s-process results

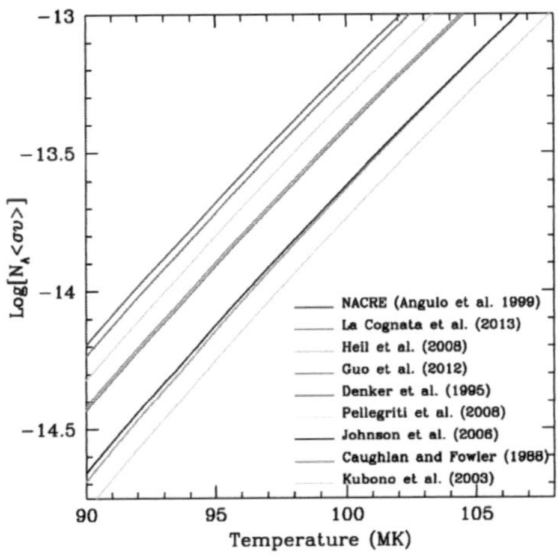

Figure 4. The rate of the ^{13}C(α,n)^{16}O reaction in the narrow range of temperatures typical of the radiative ^{13}C pocket from different literature sources: Angulo et al. (NACRE) [21]; La Cognata et al. [22]; Heil et al. [23]; Guo et al. [24]; Denker et al. [25]; Pellegriti et al. [26]; Johnson et al. [27]; Caughlan & Fowler [28]; and Kubono et al. [29].

A number of estimates are available for the rate of the ^{13}C(α,n)^{16}O reaction and are shown in Figure 4 for the narrow range of temperature where the reaction is activated in the ^{13}C pocket of AGB stars. Direct measurements of the rate have not yet been possible since the Gamow window is at 140 230 keV. Some studies [23] have extrapolated direct measurements performed at energies above 270 keV. Most other studies have employed indirect methods to evaluate the fundamental contribution of the sub-threshold resonance due to the 6.356 MeV state in ^{17}O. For clarity, I do not show the error bars in Figure 4, however, note that the last two estimates of the rate [24, 22] are outside of each other error bars (see Table 1 of [30]).

Most important is the question if these different rates and their disagreement have an impact on the *s*-process results. In the framework of the radiative ^{13}C pocket (i.e., for initial stellar masses > 2 M$_\odot$) there are no changes in the model results using any of the rates [24, 22, 30]. This is because if all the ^{13}C burns radiatively the neutron density profile is of type *a*, as shown in Figure 2. In this case, tf the ^{13}C burns faster or slower the neutron density may somewhat increase or decrease, as the timescale of burning decreases or increases, respectively. However, the area below the curve, proportional to the neutron exposure, does not change. It should be

Journal of Physics: Conference Series **703** (2016) 012003 doi:10.1088/1742-6596/703/1/012003

warned that the different rates have not been tested yet in models that include rotation. My expectation is that the value of the rate would have a much bigger impact in these models beause of the competition of the timescales between the burning and the mixing.

On the other hand, the different $^{13}C(\alpha,n)^{16}O$ rates have an impact when the ^{13}C is ingested in the convective pulse (i.e., for initial stellar masses < 2 $M_\odot$) [24]. This is because if the rate is faster more ^{13}C burns radiatively and less is ingested in the convective region, keeping the overall neutron exposure higher. Also, if the rate is faster, the ^{13}C that gets ingested in the convective pulse burns somewhat quicker than it takes for the convective region to engulf ^{14}N, and the neutron exposure and the neutron density in the pulse increase. The branching point at ^{85}Kr is very sensitive to changes in the neutron density and can be used to test the different burning conditions. This is because it is activated already at relatively low $N_n \sim 10^8$ n/cm^{-3} given the long half life of ^{85}Kr (11 yr). This branching point produces the neutron magic nucleus ^{86}Kr, whose abundance relative to ^{82}Kr is measured in meteoritic stardust grains. Interestingly, the observed $^{86}Kr/^{82}Kr$ ratio increases with grain size, which gives us the opportunity to directly link the neutron density experienced by the material inside the deepest layers of the stars to the conditions experienced during the dust formation process in the outer layers of the star. It appears that the high $^{86}Kr/^{82}Kr$ ratio observed in the large (a few μm) grains can be produced in the low-mass AGB models where the ^{13}C burns convectively [31]. However, using the highest or the lowest rate for the $^{13}C(\alpha,n)^{16}O$ reaction results in a difference of 30% in the $^{86}Kr/^{82}Kr$ ratio, which does not allow us to reach firm conclusions. Future, planned measurements of this reaction at underground laboratories like LUNA and JUNA will allow us to settle this problem.

2.2. Post-AGB stars in the Large and Small Magellanic Clouds

De Smedt et al. [32, 33, 34] and van Aarle et al. [35] derived the abundances of many neutron-capture elements from the spectra of six post-AGB stars, the direct progeny of AGB stars, in the Large and Small Magellanic Clouds. Because we know the distances of these galaxies we can derive the absolute luminosity of these stars and comparing to evolutionary models we can infer that their initial masses were between 1 to 1.5 $M_\odot$. The metallicity is also observed to be roughly 1/10th of solar. At this low metallicity the ^{13}C-pocket models predict large abundances of Pb, because there are less Fe seed nuclei to capture the neutrons, the neutron exposure is high and the overall *s*-process pattern shifts towards the third peak at Pb [7, 36]. The expected high abundances of Pb, however, are not observed in these post-AGB stars. Lugaro et al. [37] have tried to resolve this discrepancy by changing the ^{13}C-pocket profile to simulate in a simple way the possible effect of rotation, or by artificially keeping the temperature low in the ^{13}C pocket so that the ^{13}C burns in convective conditions. As described above, both cases lead to a reduction of the neutron exposure. This, in turn, leads to less of the *s*-process abundance flux to bypass the second neutron magic number at ^{138}Ba, resulting in a lower Pb abundance, as observed. However, also the abundances of all the elements between Ba and Pb, e.g., Gd, Hf, and W, are necesserily lowered, since the bottleneck at Ba is more narrow. The conclusion is that when the Pb abundance is matched, the abundances of these elements are all underpredicted. The solution to this problem may be found outside the *s*-process framework, by moving into a neutron-capture regime where the neutron density is in-between those of the *s* and the *r* processes. If the neutron-capture flux substantially moves away from the valley of β-stability, the elemental distributions can be affected in ways that are otherwise impossible for the *s*-process due to the constraints imposed by the stable magic nuclei acting as bottlenecks on its path. The existence of this *intermediate* process was first suggested by Cowan & Rose [38] and recently modelled by Herwig et al. [39]. More details can be found in the paper by M. Pignatari (these proceedings). From the observations of the six post-AGB stars in the Magellanic Clouds we derive that this process must be relatively common in AGB stars of low mass. Proton-ingestion episodes directly inside convective pulses may drive it, however, multi-dimensional, hydrodynamical models are required to address this possibility.

2.3. The other neutron source in AGB stars: $^{22}Ne(\alpha,n)^{25}Mg$

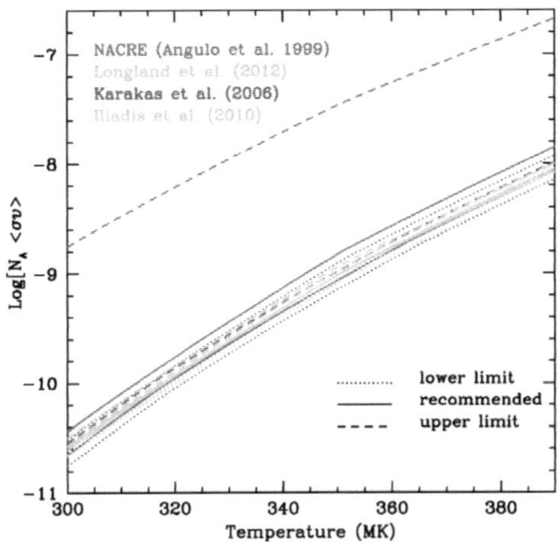

Figure 5. The rate of the ^{22}Ne(α,n)^{25}Mg in the temperature range at which it is activated in AGB stars from different literature sources, including the recommended, upper, and lower limits: Angulo et al. (NACRE) [21]; Karakas et al. [40]; Iliadis et al. [41]; and Longland et al. [42].

The ^{22}Ne(α,n)^{25}Mg reaction becomes an important neutron source only for AGB stars of masses > 3 M$_\odot$ because it requires temperatures higher than 300 MK to be activated, which are reached only in the convective pulses of these stars (Figure 3). The lowest well studied resonance at 832 keV dominates the rate [43] and lies above the relevant astrophysical energies. The influence of a possible resonance at 635 keV was included in the calculation of the upper limit of the NACRE rate (Figure 5) but then ruled out based on parity [42]. Currently, the different estimates agree with each other within the error bars (see also Table 2 of [30]), however, these theoretical extrapolations may be affected by other unknown low-energy resonances so it is not easy to eastimate the true current uncertainty. The activation of the ^{22}Ne neutron source occurs on short timescales (order of years) and results in neutron density profiles similar to case *b* of Figure 2. This means that neutron exposures are usually relatively low, favouring production of the first *s*-process peak, but neutron densities are high, even up to 10^{13} n/cm^3, so many branching points are open in these conditions. For example, the branching point at ^{86}Rb (half life 19 days) is activated by the ^{22}Ne neutron source resulting in the high Rb abundances [44] observed in massive AGB stars [45, 46]. However, the model resuts are not able to qunatitatively match the observations. As for most branching points, they are very sensitive to the ^{22}Ne(α,n)^{25}Mg rate. Only future, planned underground experiments will allow us to provide more accurate and precise model predictions.

Acknowledgments
ML is a Momentum ("Lendület-2014" Programme) project leader of the Hungarian Academy of Sciences.

References
[1] Burbidge E M, Burbidge G R, Fowler W A and Hoyle F 1957 *Rev. Mod. Phys.* **29** 547–650
[2] Dillmann I, Heil M, Käppeler F, Plag R, Rauscher T and Thielemann F 2006 KADoNiS- The Karlsruhe Astrophysical Database of Nucleosynthesis in Stars *Capture Gamma-Ray Spectroscopy and Related Topics* (*American Institute of Physics Conference Series* vol 819) ed A Woehr & A Aprahamian pp 123–127
[3] Merrill S P W 1952 *Astrophys. J.* **116** 21–26
[4] Smith V V and Lambert D L 1990 *Astrophys. J. Suppl.* **72** 387–416

[5] Abia C, Domínguez I, Gallino R, Busso M, Masera S, Straniero O, de Laverny P, Plez B and Isern J 2002 *Astrophys. J.* **579** 817–831
[6] Abia C, Busso M, Gallino R, Domínguez I, Straniero O and Isern J 2001 *Astrophys. J.* **559** 1117–1134 (*Preprint* arXiv:astro-ph/0105486)
[7] Gallino R, Arlandini C, Busso M, Lugaro M, Travaglio C, Straniero O, Chieffi A and Limongi M 1998 *Astrophys. J.* **497** 388
[8] Goriely S and Mowlavi N 2000 *å* **362** 599–614
[9] Lugaro M, Herwig F, Lattanzio J C, Gallino R and Straniero O 2003 *Astrophys. J.* **586** 1305–1319
[10] Cristallo S, Straniero O, Gallino R, Piersanti L, Domínguez I and Lederer M T 2009 *Astrophys. J.* **696** 797–820 (*Preprint* 0902.0243)
[11] Goriely S and Siess L 2004 *Astron. Astrophys.* **421** L25–L28
[12] García-Hernández D A, Zamora O, Yagüe A, Uttenthaler S, Karakas A I, Lugaro M, Ventura P and Lambert D L 2013 *Astron. Astrophys.* **555** L3 (*Preprint* 1306.2134)
[13] Herwig F, Langer N and Lugaro M 2003 *Astrophys. J.* **593** 1056–1073
[14] Siess L, Goriely S and Langer N 2004 *Astron. Astrophys.* **415** 1089–1097
[15] Piersanti L, Cristallo S and Straniero O 2013 *Astrophys. J.* **774** 98 (*Preprint* 1307.2017)
[16] Spruit H C 2002 *Astron. Astrophys.* **381** 923–932 (*Preprint* astro-ph/0108207)
[17] Mosser B, Goupil M J, Belkacem K, Marques J P, Beck P G, Bloemen S, De Ridder J, Barban C, Deheuvels S, Elsworth Y, Hekker S, Kallinger T, Ouazzani R M, Pinsonneault M, Samadi R, Stello D, García R A, Klaus T C, Li J, Mathur S and Morris R L 2012 *Astron. Astrophys.* **548** A10 (*Preprint* 1209.3336)
[18] Bradley P A 2001 *Astrophys. J.* **552** 326–339
[19] Cantiello M, Mankovich C, Bildsten L, Christensen-Dalsgaard J and Paxton B 2014 *Astrophys. J.* **788** 93 (*Preprint* 1405.1419)
[20] Lugaro M, Tagliente G, Karakas A I, Milazzo P M, Käppeler F, Davis A M and Savina M R 2014 *Astrophys. J.* **780** 95 (*Preprint* 1311.2660)
[21] Angulo C, Arnould M, Rayet M, Descouvemont P, Baye D, Leclercq-Willain C, Coc A, Barhoumi S, Aguer P, Rolfs C, Kunz R, Hammer J W, Mayer A, Paradellis T, Kossionides S, Chronidou C, Spyrou K, degl'Innocenti S, Fiorentini G, Ricci B, Zavatarelli S, Providencia C, Wolters H, Soares J, Grama C, Rahighi J, Shotter A and Lamehi Rachti M 1999 *Nucl. Phys. A* **656** 3–183
[22] La Cognata M, Spitaleri C, Trippella O, Kiss G G, Rogachev G V, Mukhamedzhanov A M, Avila M, Guardo G L, Koshchiy E, Kuchera A, Lamia L, Puglia S M R, Romano S, Santiago D and Spartà R 2013 *Astrophys. J.* **777** 143
[23] Heil M e a 2008 *Phys. Rev. C* **78** 025803–+
[24] Guo B, Li Z H, Lugaro M, Buntain J, Pang D Y, Li Y J, Su J, Yan S Q, Bai X X, Chen Y S, Fan Q W, Jin S J, Karakas A I, Li E T, Li Z C, Lian G, Liu J C, Liu X, Shi J R, Shu N C, Wang B X, Wang Y B, Zeng S and Liu W P 2012 *Astrophys. J.* **756** 193 (*Preprint* 1208.0714)
[25] Denker A, Drotleff H W, Grosse M, Knee H, Kunz R, Mayer A, Seidel R, Soiné M, Wöohr A, Wolf G and Hammer J W 1995 Neutron Producing Reactions in Stars *Nuclei in the Cosmos III* (*American Institute of Physics Conference Series* vol 327) ed Busso M, Raiteri C M and Gallino R p 255
[26] Pellegriti M G, Hammache F, Roussel P, Audouin L, Beaumel D, Descouvemont P, Fortier S, Gaudefroy L, Kiener J, Lefebvre-Schuhl A, Stanoiu M, Tatischeff V and Vilmay M 2008 *Phys. Rev. C* **77** 042801
[27] Johnson E D, Rogachev G V, Mukhamedzhanov A M, Baby L T, Brown S, Cluff W T, Crisp A M, Diffenderfer E, Goldberg V Z, Green B W, Hinners T, Hoffman C R, Kemper K W, Momotyuk O, Peplowski P, Pipidis A, Reynolds R and Roeder B T 2006 *Physical Review Letters* **97** 192701
[28] Caughlan G R and Fowler W A 1988 *Atomic Data and Nuclear Data Tables* **40** 283
[29] Kubono S, Abe K, Kato S, Teranishi T, Kurokawa M, Liu X, Imai N, Kumagai K, Strasser P, Tanaka M H, Fuchi Y, Lee C S, Kwon Y K, Lee L, Ha J H and Kim Y K 2003 *Physical Review Letters* **90** 062501
[30] Bisterzo S, Gallino R, Käppeler F, Wiescher M, Imbriani G, Straniero O, Cristallo S, Görres J and deBoer R J 2015 *Mon. Not. R. Astron. Soc.* **449** 506–527 (*Preprint* 1507.06798)
[31] Raut R, Tonchev A P, Rusev G, Tornow W, Iliadis C, Lugaro M, Buntain J, Goriely S, Kelley J H, Schwengner R, Banu A and Tsoneva N 2013 *Physical Review Letters* **111** 112501 (*Preprint* 1309.4159)
[32] De Smedt K, Van Winckel H, Karakas A I, Siess L, Goriely S and Wood P R 2012 *Astron. Astrophys.* **541** A67 (*Preprint* 1203.4413)
[33] De Smedt K, Van Winckel H, Kamath D, Karakas A I, Siess L, Goriely S and Wood P 2014 *Astron. Astrophys.* **563** L5 (*Preprint* 1402.3169)
[34] De Smedt K, Van Winckel H, Kamath D and Wood P R 2015 *Astron. Astrophys.* **583** A56 (*Preprint* 1508.01091)
[35] van Aarle E, Van Winckel H, De Smedt K, Kamath D and Wood P R 2013 *Astron. Astrophys.* **554** A106 (*Preprint* 1304.7103)

[36] Travaglio C, Gallino R, Busso M and Gratton R 2001 *Astrophys. J.* **549** 346–352 (*Preprint* astro-ph/0011050)
[37] Lugaro M, Campbell S W, Van Winckel H, De Smedt K, Karakas A I and Käppeler F 2015 *Astron. Astrophys.* **583** A77 (*Preprint* 1509.03518)
[38] Cowan J J and Rose W K 1977 *Astrophys. J.* **212** 149–158
[39] Herwig F, Pignatari M, Woodward P R, Porter D H, Rockefeller G, Fryer C L, Bennett M and Hirschi R 2011 *Astrophys. J.* **727** 89 (*Preprint* 1002.2241)
[40] Karakas A I, Lugaro M A, Wiescher M, Görres J and Ugalde C 2006 *Astrophys. J.* **643** 471–483 (*Preprint* astro-ph/0601645)
[41] Iliadis C, Longland R, Champagne A E and Coc A 2010 *Nuclear Physics A* **841** 323–388 (*Preprint* 1004.4150)
[42] Longland R, Iliadis C and Karakas A I 2012 *Phys. Rev. C* **85** 065809 (*Preprint* 1206.3871)
[43] Drotleff H W, Denker A, Knee H, Soine M, Wolf G, Hammer J W, Greife U, Rolfs C and Trautvetter H P 1993 *Astrophys. J.* **414** 735–739
[44] van Raai M A, Lugaro M, Karakas A I, García-Hernández D A and Yong D 2012 *Astron. Astrophys.* **540** A44 (*Preprint* 1202.2620)
[45] Garcia-Hernandez D A, Garcia-Lario P, Plez B, D'Antona F, Manchado A and Trigo-Rodriguez J M 2006 *Science* **314** 1751 (*Preprint* astro-ph/0611319)
[46] Zamora O, García-Hernández D A, Plez B and Manchado A 2014 *Astron. Astrophys.* **564** L4 (*Preprint* 1403.2075)

Journal of Physics: Conference Series **703** (2016) 012004 doi:10.1088/1742-6596/703/1/012004

Introduction to Galactic Chemical Evolution

Francesca Matteucci

Department of Physics, Trieste University, Via G.B. Tiepolo 11, 34131 Trieste, Italy

E-mail: matteucci@oats.inaf.it

Abstract. In this lecture I will introduce the concept of galactic chemical evolution, namely the study of how and where the chemical elements formed and how they were distributed in the stars and gas in galaxies. The main ingredients to build models of galactic chemical evolution will be described. They include: initial conditions, star formation history, stellar nucleosynthesis and gas flows in and out of galaxies. Then some simple analytical models and their solutions will be discussed together with the main criticisms associated to them. The yield per stellar generation will be defined and the hypothesis of instantaneous recycling approximation will be critically discussed. Detailed numerical models of chemical evolution of galaxies of different morphological type, able to follow the time evolution of the abundances of single elements, will be discussed and their predictions will be compared to observational data. The comparisons will include stellar abundances as well as interstellar medium ones, measured in galaxies. I will show how, from these comparisons, one can derive important constraints on stellar nucleosynthesis and galaxy formation mechanisms. Most of the concepts described in this lecture can be found in the monograph by Matteucci (2012).

1. Introduction

We call chemical evolution of galaxies the study of how the chemical elements formed in stars and were distributed in galaxies. During the Big Bang only the light elements (H, D, He, Li) were synthesized, while stars have been responsible for the formation and distribution of all the elements from carbon to uranium and beyond. Some light elements, such as Li, Be and B are formed during the spallation process, which is the interaction between the cosmic rays and C,N,O atoms present in the interstellar medium (ISM). Therefore, stars produce chemical elements in their interiors by means of nuclear reactions of fusion and then restore these elements into the ISM when they die. The fusion reactions occur up to ^{56}Fe which is the element with the maximum binding energy per nucleon. For elements heavier than ^{56}Fe the nuclear fusion is therefore inhibited and the nuclear fission is favored. The stars are born, live and die and they can die in a quiescent fashion like white dwarfs or violently as supernovae (SNe). The supernovae can be of various types: core-collapse SNe, namely the explosion of single massive stars ($M \geq 10M_{\odot}$, Type II, Ib, Ic) , and Type Ia SNe, namely the explosion of a white dwarf occurring after accreting material from a companion in a binary system. In this lecture I will first describe the ingredients necessary to build galactic chemical evolution models. Such models can be analytical or numerical and they aim at following the evolution in time and space of the abundances of the chemical elements in the ISM. Then, I will focus on some highlights in the galactic chemical evolution and on the comparison models-observations. In doing so, I will show how chemical evolution models can constrain stellar nucleosynthesis and galaxy formation timescales.

Journal of Physics: Conference Series **703** (2016) 012004 doi:10.1088/1742-6596/703/1/012004

2. Basic ingredients for galactic chemical evolution

The basic ingredients necessary to build a chemical evolution model are:

- **Initial conditions**
  The initial conditions consist in deciding whether to assume an open or a closed model, in other words to decide if the gas, out of which the stars form, is present since the beginning or if it is accreted during the galactic lifetime; moreover, one should assume whether the initial gas has a primordial (only light elements from the Big Bang) or enriched chemical composition.
- **Stellar birthrate function (SFRxIMF)**
  The stellar birthrate function is the history of star formation in a galaxy and it can be expressed as the product of the star formation rate (SFR) per the initial mass function (IMF), namely:

$$B(m,t) = \psi(t) \cdot \phi(m), \tag{1}$$

  where the function $\psi(t)$ represents the SFR and it is generally assumed to be only a function of time, whereas $\phi(m)$ is the IMF which is assumed to be only a function of mass. The SFR represents how many solar masses go into stars per unit time, while the IMF is the initial stellar mass function describing the distribution of stars at birth as a function of the stellar mass. Clearly, these hypotheses are semplifications and we do not know whether the SFR is independent of mass and the IMF independent of time.

2.1. Parametrization of the SFR

The most common parametrization is the Schmidt (1959) law where the SFR is proportional to some power k of the gas density. Kennicutt (1998) suggested $k = 1.4 \pm 0.15$, as deduced by data relative to star forming galaxies. Other important parameters such as gas temperature, viscosity and magnetic field are usually ignored.
The Kennicutt SFR can be written as:

$$\psi(t) = \nu \sigma^k_{gas}, \tag{2}$$

where σ^k_{gas} is the gas surface mass density and ν is the efficiency of star formation, namely the SFR per unit mass of gas.

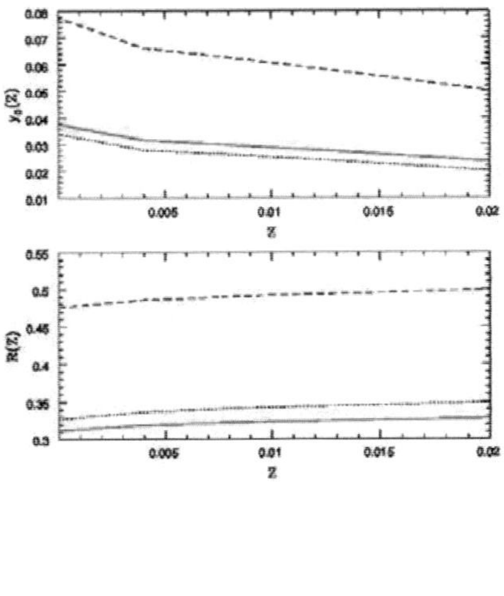

Figure 1. Upper panel: the yield of oxygen per stellar generation computed for different metallicities and IMFs The blue dotted line refers to Chabrier (2003) IMF, the magenta line is the Salpeter (1955) IMF and the black line is the Kroupa et al. (1993) IMF. Lower panel:returned fraction R as a function of IMF and metallicity. Figure from Vincenzo et al. (2015).

2.2. Parametrization of the IMF
The IMF, namely the number of stars born in the mass interval, m-m+dm, is generally expressed as a power law. It is possible to measure the IMF only in the solar vicinity since one needs to count the stars as functions of their magnitudes and it is still not possible to do it in external galaxies. Therefore, the only observational information we have is relative to the solar region of our Galaxy.

- **Stellar nucleosynthesis**
The stellar yields are defined as the masses of chemical elements produced by stars of different masses. These yields can represent both the newly formed elements or the elements already present in the star at its formation and restored into the ISM without being reprocessed. In particular, we can define the yield of a newly formed element, p_{im}, as the integral over the stellar lifetime of the mass loss rate multiplied by [X(i)-Xo(i)], where Xo(i) is the original abundance of the element i. To obtain the total stellar yield, it should be added to this term the mass ejected without processing, namely Xo(i) x mass lost, where with mass lost we intend the total mass ejected by a star into the ISM during its lifetime. Each stellar mass can produce and eject different chemical elements and the yields are therefore a function of the stellar mass but also of the original stellar metal content that we will indicate with Z. These yields are computed by means of detailed nucleosynthesis calculations taking into account all the main nuclear reactions in stars. Here we summarize briefly the element production in stars: i) low and intermediate mass stars (0.8-8 $M_{\odot}$): produce He, N, C and heavy s-process elements. They die as C-O white dwarfs, when single, and can die as Type Ia SNe when binaries. ii) Stars with $M < 0.8M_{\odot}$ do not contribute to galactic chemical enrichment and have lifetimes longer than the Hubble time. iii) Massive stars ($M > 8 - 10M_{\odot}$): they produce mainly alpha-elements (O, Ne, Mg, S, Si, Ca), some Fe, light s-process elements and perhaps r-process elements (?) and explode as core-collapse SNe. However, r-process elements originating in neutron binary mergers seems to represent the most promising channel for r-process element production these days (see Matteucci et al, 2014 and references therein).
- **Gas flows: infall, outflow, radial flows**
In order to build a realistic galaxy one has to assume the presence of gas flows both in and out. The gas inflows are considered either as gas accretion or radial gas flows and they are influencing the chemical evolution of galaxies: in the case of accretion , usually assumed at a constant rate or exponentially decreasing in time, the main effect is to dilute the metal content. In the case of outflows or galactic winds also the effect is to decrease the metal concentration by decreasing the gas which is available for star formation. Galactic outflows are generally assumed to occur at a rate proportional to the star formation rate.

3. The yield per stellar generation
The "yield per stellar generation" of a single chemical element, can be defined as (Tinsley 1980):

$$y_i = \frac{\int_1^{\infty} m p_{im} \phi(m) dm}{(1 - R)} \tag{3}$$

where p_{im} is the mass of the newly produced element i ejected by a star of mass m, and R is the returned fraction. The yield y_i is therefore the mass fraction of the element i newly produced by a generation of stars relative to the fraction of mass in remnants (white dwarfs, neutron stars and black holes) and never dying low mass stars ($M < 0.1M_{\odot}$).

We define "returned fraction" the fraction of mass ejected into the ISM by an entire stellar generation, namely:

$$R = \int_1^{\infty} (m - M_{rem}) \phi(m) dm \tag{4}$$

The term fraction originates from the fact that both y_i and R are divided by the normalization condition of the IMF, namely:

$$\int_{0.1}^{\infty} m\phi(m)dm = 1. \quad (5)$$

In order to define y_i and R we have made a very specific assumption: the instantaneous recycling approximation (IRA), which states that *all stars more massive than 1$M_\odot$ die instantaneously, while all stars less massive live forever.* This assumption allows us to solve the chemical evolution equations analytically, as we will see in the following, but it is a very poor approximation for chemical elements produced on long timescales such as C, N and Fe. On the other hand, for oxygen, which is almost entirely produced by short lived core-collapse SNe, IRA is an acceptable approximation. In Figure 1 we show y_O and R, computed for different initial metallicities of the stars and different IMFs. As one can see, the dependence of these quantities on Z is negligible whereas that on the IMF is strong. In that Figure it does not appear the Kroupa(2001) universal IMF, suggesting that the IMF in stellar clusters is an universal one. Kroupa (2001) IMF is a two-slope IMF with a slope for stars more massive than $0.5M_\odot$ very similar to that of the Salpeter (1955) IMF, still widely used in model for external galaxies.

4. The Simple Model

The Simple Model of chemical evolution assumes that the system is evolving as a closed-box, without inflows or outflows, the IMF is constant in time, the chemical composition of the gas is primordial and the mixing between the chemical products ejected by stars and the ISM is instantaneous, plus IRA.

The initial gas mass is therefore, $M_{gas}(0) = M_{tot} = constant$, $\mu = M_{gas}/M_{tot}$ is the fractionary mass of gas, the metallicity is $Z = M_Z/M_{gas}$ is zero at the time t=0. The basic equations can be written as:

$$\frac{dM_{gas}}{dt} = -\psi(t) + \int_{m(t)}^{\infty} (m - M_R)\psi(t - \tau_m)\varphi(m)dm \quad (6)$$

which describes the evolution of gas and where the integral is the rate at which dying stars restore both the enriched and unenriched material into the ISM at the time t.

The equation for metals is:

$$\frac{d(ZM_{gas})}{dt} = -Z\psi(t) + \int_{m(t)}^{\infty} [(m - M_R)Z(t - \tau_m) + mp_{Zm}]\psi(t - \tau_m)\varphi(m)dm, \quad (7)$$

where the first term in the square brackets represents the mass of pristine metals which are restored into the ISM without suffering any nuclear processing, whereas the second term contains the newly formed and ejected metals (Maeder 1992). When IRA is assumed, the SFR can be taken out of the integrals and the equation for metals be solved analytically. Its solution is:

$$Z = y_Z ln(\frac{1}{\mu}). \quad (8)$$

The metallicity yield per stellar generation y_Z which appears in the above equation is known as *effective yield*, simply defined as the yield $y_{Z_{eff}}$ that would be deduced if the system were assumed to be described by the Simple Model.

Therefore, the effective yield is:

$$y_{Z_{eff}} = \frac{Z}{ln(1/\mu)}. \quad (9)$$

Clearly, the true yield y_Z will be always lower than the effective one in both cases of winds and infall of primordial gas. The only way to increase the effective yield is to assume an IMF more weighted towards massive stars than the canonical IMFs.

4.1. Primary and secondary elements

We define primary element an element produced directly from H and He A typical primary element is carbon or oxygen which originate from the 3-α reactions We define secondary element an element produced starting from metals already present in the star at birth (e.g. nitrogen produced in the CNO cycle). We recall that the solution of the chemical evolution equations for a secondary element, with abundance X_S, implies:

$$X_S = \frac{1}{2}(\frac{y_S}{y_Z Z_{\odot}})Z^2, \tag{10}$$

where y_S is the yield per stellar generation of the secondary element. This means that, for a secondary element the simple closed-box model predicts that its abundance increases proportionally to the metallicity squared, namely:

$$X_S \propto Z^2. \tag{11}$$

5. Analytical solutions for outflow and inflow

In the case of a model with outflow but no inflow occurring at a rate:

$$W(t) = \lambda(1 - R)\psi(t), \tag{12}$$

where λ is a free parameter larger than or equal to zero, the solution for the metallicity of the system is:

$$Z = \frac{y_Z}{(1+\lambda)} ln[(1+\lambda)\mu^{-1} - \lambda]. \tag{13}$$

It is clear that in the case of $\lambda = 0$ the solution is the same as that of the Simple Model.

In the case of a model without outflow but inflow of primordial gas ($Z_{inf} = 0$), occurring at a rate:

$$A(t) = \Lambda(1 - R)\psi(t) \tag{14}$$

with Λ a positive constant different from zero and from 1, the solution is:

$$Z = \frac{y_Z}{\Lambda}[1 - (\Lambda - (\Lambda - 1)\mu^{-1})^{-\Lambda/(1-\Lambda)}]. \tag{15}$$

Again, if $\Lambda = 0$, the solution is the same as that of the Simple Model. The case $\Lambda = 1$ is a particular one, called "extreme infall" and it has a different solution:

$$Z = y_Z(1 - e^{-\beta}), \tag{16}$$

where $\beta = \mu^{-1} - 1$. For a more extensive discussion of these and other analytical solutions see Matteucci (2012).

6. Numerical models

When the stellar lifetimes are correctly taken into account the chemical evolution equations should be solved numerically. This allows us to follow in detail the temporal evolution of the abundances of single elements.

A complete chemical evolution model in the presence of both galactic wind, gas infall and radial flows can be described by a number of equations equal to the number of chemical species: in particular, if M_i is the mass of the gas in the form of any chemical element i, we can write the following set of integer-differential equations which can be solved only numerically, if IRA is relaxed:

$$\dot{M}_i(t) = -\psi(t)X_i(t) + \int_{M_L}^{M_{Bm}} \psi(t-\tau_m)Q_{mi}(t-\tau_m)\varphi(m)dm + \tag{17}$$
$$A_B \int_{M_{Bm}}^{M_{BM}} \varphi(m)[\int_{\mu_{B_{min}}}^{0.5} f(\mu_B)\psi(t-\tau_{m2})Q_{mi}(t-\tau_{m2})d\mu_B]dm +$$
$$(1-A_B)\int_{M_{Bm}}^{M_{BM}} \psi(t-\tau_m)Q_{mi}(t-\tau_{m})\varphi(m)dm +$$
$$\int_{M_{BM}}^{M_U} \psi(t-\tau_m)Q_{mi}(t-\tau_m)\varphi(m)dm + X_{iA}(t)A(t)$$
$$-X_i(t)W(t) + X_i(t)I(t),$$

where M_i can be substituted by σ_i, namely the surface gas density of the element i. In several models of chemical evolution it is customarily to use normalized variables which should be substituted to $M_i(t)$ or to $\sigma_i(t)$, such as for example:

$$G_i(t) = \frac{\sigma_i(t)}{\sigma_{tot}(t_G)} = \frac{M_i(t)}{M_{tot}(t_G)}, \tag{18}$$

with $\sigma_i(t) = X_i(t)\sigma_{gas}(t)$, and $\sigma_{tot}(t_G)$ ($M_{tot}(t_G)$) being the total surface mass density (mass) at the present time t_G. The surface densities are more indicated for computing the chemical evolution of galactic disks, while for spheroids one can use the masses. The quantity $X_i(t) = \frac{\sigma_i(t)}{\sigma_{gas}(t)} = \frac{M_i(t)}{M_{gas}(t)}$ represents the abundance by mass of the element i and by definition the summation over all the mass abundances of the elements present in the gas mixture is equal to unity. These equations include the products from Type Ia SNe (third term on the right) and the products of stars ending their lives as white dwarfs and core-collapse SNe (for a more extensive description of the equations see Matteucci, 2012).

7. Some highlights

In this section we will illustrate some examples where predictions from chemical evolution models are compared to observations.

7.1. The chemical evolution of the Milky Way

We will start with the Milky Way, which is the galaxy for which we possess the majority of information. A good model for the chemical evolution of the Milky Way should reproduce several constraints, including the G-dwarf metallicity distribution, the [X/Fe] vs. [Fe/H][1] relations (where X is a generic element from carbon up to the heaviest ones), abundance and gas gradients along the Galactic disk, among others.

In Figure 2 we show the predictions from Romano et al. (2010) concerning several chemical elements obtained by using different sets of stellar yields and compared to observation in stars. As one can see, the agreement is good for some chemical species whereas for others the agreement is still very poor. The reason for that probably resides in the uncertainties still existing in the theoretical stellar yields. The shapes of the [X/Fe] vs. [Fe/H] relation can be successfully interpreted as due to the *time-delay model*, namely the fact that elements such as C, N and Fe are mainly produced by long living stars, while others such as α-elements (O, Mg, Si, Ca) are produced by short living stars. In particular, Fe is mainly produced by Type Ia SNe and only a small fraction of it originates in core-collapse (CC) SNe. Type I a SNe explode on longer

[1] [Fe/H]= $log(Fe/H)_* - log(Fe/H)_\odot$

timescales than core-collapse SNe and therefore ratios such as [α/Fe] , where α indicates α-elements which are mainly produced in CC SNe, can be used as cosmic clocks. The higher than solar value of the [α/Fe] ratios for low [Fe/H] values, is then due to the CC SNe which restore the α-elements on short timescales. When Type Ia SNe, originating from CO white dwarfs which have longer lifetimes (from 30 Myr up to a Hubble time), start restoring the bulk of Fe, then the [α/Fe] ratios start decreasing. By means of the time-delay model we can interpret any abundace ratio.

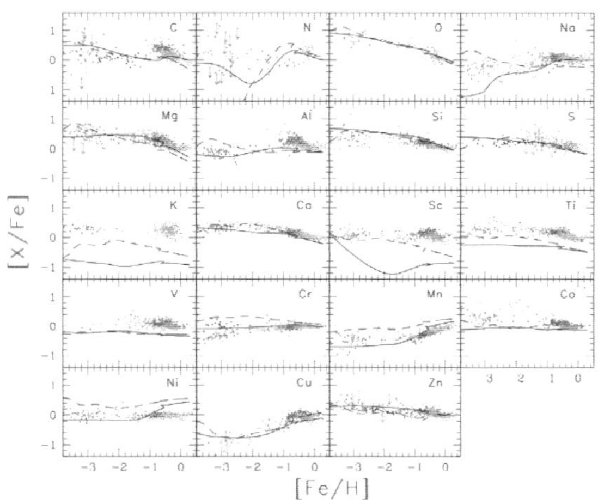

Figure 2. Comparison between model predictions and observatioons. The two sets of models refer to different stellar yields: the dashed lines refer to the yields of Woosley & Weaver (1995) for massive stars and van den Hoek and Groenewegen (1997) for low and intermediate mass stars, whereas the continuous lines represent the yields of Kobayashi et al. (2006) for massive stars, Karakas (2010) for low and intermediate masses and for CNO elements the yields of the Geneva group with mass loss and rotation. Figure from Romano et al. (2010).

The model underlying the predictions of Figure 2 is an updated version of the two-infall model by Chiappini et al. (1997): this model assumes that the Milky Way formed by means of two main gas accretion episodes, one during which the halo and thick disk formed and another during which the thin disk formed on much longer timescales.

7.2. The Li-problem

In Figure 3 we show an illustration of the so-called Li-problem. In particular, we report the abundance of ^7Li (A(Li) = 12 +log(Li/H) versus [Fe/H]). The data sources are indicated inside the Figure. The upper envelope of the data should represent the evolution of the abundance of ^7Li in the ISM since the stars in the upper envelope should exhibit the Li abundance present in the gas when they formed. The stars below should instead have consumed the original Li which is very fragile and tends to react with protons to form ^4He when the temperature is $> 2.8 \cdot 10^6$ K. According to this interpretation of the diagram, the younger stars with higher [Fe/H] should reflect the fact that the Li abundance has increased with galactic lifetime owing to Li stellar production by stars. Stars which can create, preserve and eject Li into the ISM are: low and intermediate mass stars during the asymptotic giant branch (AGB) phase, the CC SNe, cosmic rays and perhaps novae (see Izzo et al. 2015 for the possible detection of Li in a nova). The Li problem arises from the fact that both WMAP and Planck experiments have derived a primordial ^7Li abundance higher than the abundance found in the upper envelope of the Galactic halo stars (those with [Fe/H]< -1.0 in Figure 3). This could mean that the oldest Galactic stars have consumed their primordial Li and roughly by the same amount for stars in the range -3.0 — -1.0 in [Fe/H] (the so-called Spite plateau). The lines in Figure 3 represent the predictions of a chemical evolution model (the two-infall model) including SNe, novae, cosmic rays and AGB stars as Li producers and starting from the primordial Li abundance deduced either by WMAP or by the value of the Spite plateau. There is not yet a solution for the Li-problem.

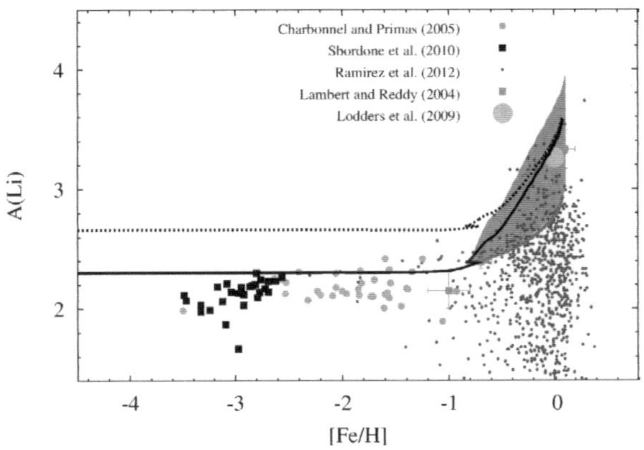

Figure 3. A(Li) vs. [Fe/H] for solar vicinity stars and meteorites (symbols; see legend) compared to the predictions of chemical evolution models (lines and colored areas). The chemical evolution model is from Romano et al. (2010).The black line represents all the stellar Li producers, as described in the text. The red line is the same model but without novae. Note that the dotted theoretical curve starts with the primordial Li value from WMAP which lies well above the Li abundances of halo stars.Figure from Izzo et al. (2015).

7.3. The Ultra-Faint Dwarfs (UFDs)

In Figure 4 we show the model predictions and data for the UFD galaxy Hercules, a small (with stellar mass of $\sim 10^7 M_\odot$) satellite of the Milky Way. The data refer to the [Ca/Fe] ratios measured in the stars of this galaxy and include also the data for Ca in the Milky Way and dwarf spheroidal stars. The data show that the Milky Way and Hercules evolved in a different way, since the [Ca/Fe] ratios are lower at the same [Fe/H] in the UFD galaxy, relative to the Milky Way. The models, computed with an efficiency of star formation one thousand times lower than assumed for the Milky Way can well reproduce the trend. The model for Hercules includes also a strong galactic wind proportional to the SFR. Because of the already mentioned time-delay model, the Fe delayed production coupled with a regime of slow star formation predict that the [Ca/Fe] ratios are lower. at the same [Fe/H], since when the Type Ia SNe start to be important in producing Fe, the Ca and the other metals produced by CC SNe have not yet attained the same abundances as in the Miilky Way which evolves with a much faster star formation. Therefore, in order to interpret the abundance patterns in galaxies we should consider the history of star formation, since it affects strongly the shape of the [X/Fe] vs. [Fe/H] relations. In this way, we can reconstruct the history of stars formation in galaxies just by looking at their abundances. This approach is known as *astroarchaeological approach*. From Figure 4 it apprears that the dwarf spheroidals have a higher star formation efficiency than the UFDs but a lower one than the Milky Way. From this Figure it is difficult to conclude that stars of the UFDs could have been the building blocks of the Galactic halo, but more data on UFDs are necessary before drawing firm conclusions on this point.

7.4. Acknowledgments

I acknowledge financial support from the PRIN2010-2011 project "The Chemical and Dynamical Evolution of the Milky Way and Local Group Galaxies", prot.2010LY5N2T.

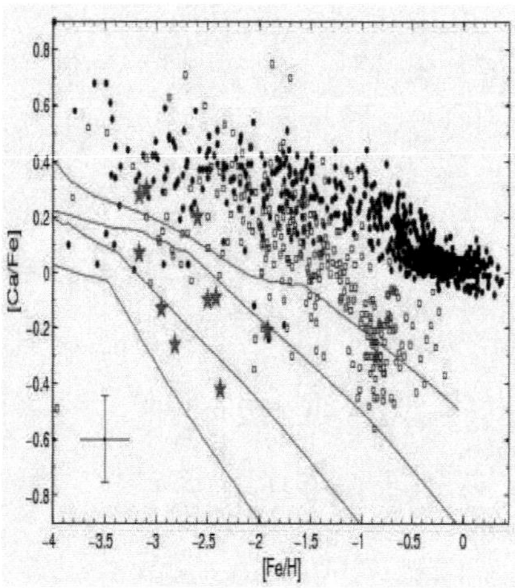

Figure 4. Theoretical and observed [Ca/Fe] ratios as functions of [Fe/H]. The red stars represent abundances in stars of Hercules, while the blue points represent dwarf spheroidals and black points the Milky Way. The red lines are models assuming a star formation efficiency (ν) varying in the range 0.002-0.008 Gyr^{-1}. Figure from Koch et al. (2012).

References

[1] Bonifacio, P. et al. 2007 *A&A* **470** 153
[2] Chabrier, G. 2003 *PASP* **115** 763)
[3] Charbonnel, C. & Primas, F. 2005 *A&A* **442** 961
[4] Karakas, A.I. 2010 *MNRAS* **403** 1413
[5] Kobayashi, C. et al. 2006 *ApJ* **653** 1145
[6] Koch, A., Matteucci, F., & Feltzing, S. 2012 *American Institute of Physics Conference Series* **1480** 190
[7] Kroupa, P: et al. 1993 *MNRAS* **262** 545
[8] Kroupa, P. 2001 *MNRAS* **322** 231
[9] Izzo et al. 2015 *ApJL* **808** L14
[10] Lambert, D.L. & Reddy, B.E. 2004 *MNRAS* **349** 757
[11] Lodders et al. 2009 *Landolt Börnstein* **44**
[12] Maeder, A. 1992 *A&A* **264** 105
[13] Matteucci, F. 2012 "Chemical Evolution of Galaxies" *Astronomy and Astrophysics Library.Springer-Verlag Berlin Heidelberg* 2012
[14] Matteucci, F. et al. 2014 *MNRAS* **438** 3177
[15] Ramirez, I. et al. 2012 *ApJ* **756** 46
[16] Romano et al. 2010 A&A **522** 32
[17] Salpeter, E.E. 1955 *ApJ* **121** 161
[18] Sbordone et al. 2010 *A&A* **522** 26
[19] Tinsley, B.M. 1980 *Fund. Cosmic Physics* **5** 287
[20] van den Hoeck, L.B. & Groenewegen, M.A.T. 1997 *A&ASuppl* **123** 305
[21] Vincenzo, F. et al. 2015 *arXiv:1503.08300*
[22] Woosley, S.E. & Weaver, T.A. 1995 *ApJS* **101** 181

Journal of Physics: Conference Series **703** (2016) 012005 doi:10.1088/1742-6596/703/1/012005

Introduction to Asymptotic Giant Branch Stars

Mounib F. El Eid

American niversity of Beirut, epartment of hysics, .O. Box 11 236 Riad l Solh, Beirut ebanon

mail: meid@aub.edu.lb

Abstract. A brief introduction on the main characteristics of the asymptotic giant branch stars (briefly: AGB) is presented. e describe a lin to observations and outline basic features of theoretical modeling of these important evolutionary phases of stars.The most important aspects of the AGB stars is not only because they are the progenitors of white dwarfs, but also they represent the site of almost half of the heavy element formation beyond iron in the galaxy. These elements and their isotopes are produced by the s process nucleosynthesis, which is a neutron capture process competing with the β^- radioactive decay.The neutron source is mainly due to the reaction $^{13}C(\alpha,n)^{16}O$ reaction. It is still a challenging problem to obtain the right amount of ^{13}C that can lead to s process abundances compatible with observation. Some ideas are presented in this context.

1. Introduction

bservations of M type stars show their spectra dominated by molecular spectra, such as Ti . The observations of carbon stars show that the > 1 in contrast to normal stars or interstellar medium.In other words, these stars are carbon rich due to the third dredge up during the thermal pulsations of AGB stars described below in sect.3.

Another spectral peculiarity of cool giants is enhanced bands of s process elements, in particular r , which again indicates the role of the third dredge up. Note that stars showing bands of r and Ti are called MS stars. An interesting element produced in AGB stars is ^{19}F by neutron capture process. A famous element usually uoted is the Technetium (^{99}Tc) with halflife time of $2 \cdot 1$ ^{5} yrs. This s process element serves as direct observational evidence of a new heavy element produced in a star. More about the s process in the extended tal s by M. ucaro in this proceeding. The luminosity can reach values more than 1 , solar luminosity and bolometric magnitudes is in the range M_{bol} $-3.6 - 7.61$.

The R diagram of globular clusters showing the existence of AGB stars is a fundamental demonstration of their formation. ue to the limited space for this contribution,we only summari e some important issues related the AGB stars.Their evolutionary route is described in sect.2., the main characteristics of the AGB evolution is given in sect.3, concluding remar s are presented in sect.4.

2. The evolution to the AGB stage

very star formed in the universe must evolve through the phase of core hydrogen burning either by the proton proton chain for stars of masses M$\leq$ 2. M$_\odot$, or by the N cycle for higher masses. This evolutionary phase has deep meaning. It is characteri ed by the low e ciency

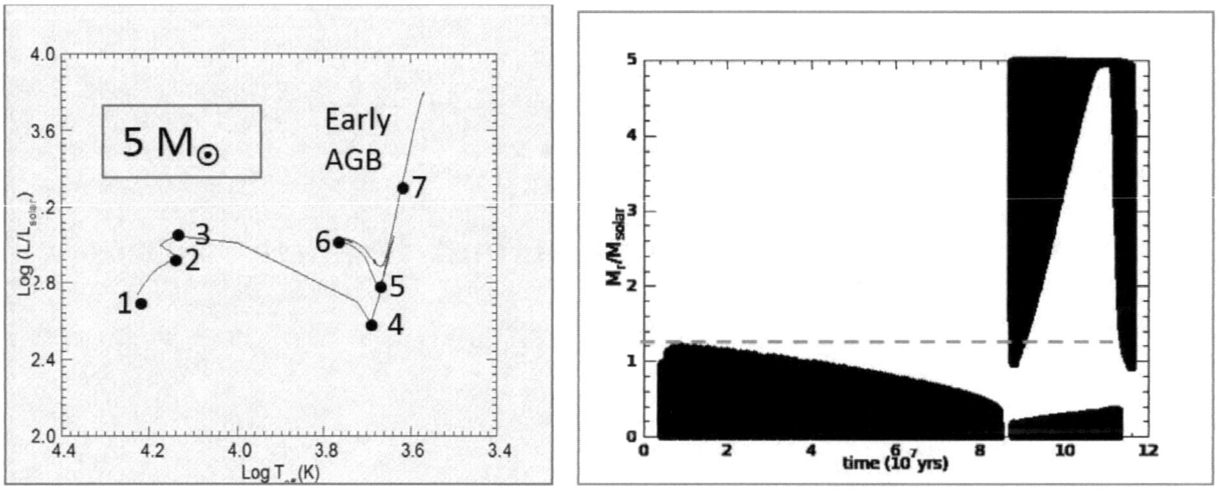

Figure 1. volution of a $5\,M_\odot$ star of solar li e initial composition in the R diagram (left panel), and its internal structural evolution with time (right panel). Note that the blac regions are convective. This calculation was carried up to the early AGB stars. The rst extension of the outer convective envelope mar s the *first dredge up*, and the send downward penetration mar s the *second dredge up*. etails of such calculations including other masses are described in 4

of nuclear burning, namely the rate of transformation of to e of .71%. This is the mass di erence in synthesi ing helium from four protons. This number is one of six numbers shaping the universe 1 . The main characteristics of the evolution of stars which reach the AGB stage is as follows:

- The relevant mass range for this type of evolution is $(1-1\ )M\,M_\odot$. It is a real challenging tas to discuss this mass range. ne distinguish between the normal AGB stars which are the progenitors of the carbon oxygen (or) white dwarfs, which do not evolve though the carbon burning phase. A review paper on AGB stars is given by 2 . The so called super AGB stars are the massive ones, which evolve through carbon burning to form electron degenerate oxygen Neon (or Ne) white dwarfs. The stars which do not end as Ne white dwarfs belong to the electron capture supernovae. It is rather surprising to read in a very recent preprint 14 the following remar s: the initial mass range for entering the super AGB phase is $(7-\ )\,M_\odot$ as obtained by 5 with assumed initial solar metallicity, but the critical mass for carbon ignition could decrease to $5.4\,M_\odot$ according to 6 . The reason for this strong variation is attributed to the treatment of the convective boundary mixing and to the yet uncertain 12 12 reaction rate. In any case, the mass range representing the progenitors of the Ne white dwarfs seems to be rather narrow.
- The inspection of Fig.1 reveals the evolutionary phases till the early AGB stage. The main se uence phase occurs between the positions 1 to 2 corresponding to core hydrogen burning occurring in a convective core of decreasing mass as shown in the right panel of Fig.1, where hydrogen transformed to helium. This phase is followed by shell hydrogen burning (burning). The released energy leads the star to evolve to the red giant branch (RGB), from positions mar ed 3 to 4. The expansion leads to an increase of the opacity causing the enveloper to become convective and start penetrating the stellar layers to mix the product

of shell burning to the surface as seen in the right panel of Fig. 1 (details of this sub ect is found in a recent wor by 3 . this stage is mar ed as *first dredge up*(F . ore e burning starts at position 4 in the Fig. 1 and the star ascends the RGB. The core e burning diminishes the role of the burning shell, so that the luminosity drops to position 5 in Fig. 1. But what happens subse uently between positions 5 and 6 is the occurrence of a loop to higher e ective temperature, the so called blue loop triggered by the increasing energy production in the hydrogen burning shell encountering higher hydrogen abundance created by the F . This episode is described in details in the paper by (4 , 3). The ma or part of core e burning occurs in the neighborhood of position 6 in Fig. 1. The evolution proceeds bac to the RGB, where a second *second dredge up occurs* (S), before the star evolves to position 7 mar ed as early AGB stage. It is emphasi ed that the S occurs only in stars of masses $\geq 4\,M_{\odot}$, since the envelope convection does not extend to the processed layers by the burning shell in these relatively lower mass stars.

- hen the star reaches the *early AGB stage*(about position 7 in Fig. 1), it has the structure consisting of a core, a helium shell and a hydrogen shell. The luminosity of the star is mainly supported by the burning shell, which becomes wea er due to the expanding outer layers. This leads the star to contract until the e rich thin layers become hot enough to ignite e in a thin stellar region. The large energy release by the triple alpha process characteri ing the e burning cannot be transported by radiation with the result that a helium convective one develops and forces the star to expand. This mar s the thermal pulsation of the AGB phase of evolution. In other words, thermal pulsation is an a double shell burning with short (order of years) rather violent e flash and a long phase(greater than 3 , years) of shell hydrogen burning. In the next sect., we give some comments on the AGB evolution describing some open problems.

3. Remarks on the AGB evolution

This evolutionary phase is the most complicated in this context due to the many involved physical e ects (description of convective mixing and burning, mass loss, e uation of state, opacities and numerical treatment). e use Fig. 2 to describe the main characteristics of this type of evolution. This Fig. shows two consecutive thermal pulses in a schematic way. After e ignition in the thin e shell and the formation of the convective e burning shell, the convective envelope recedes in mass and the burning shell becomes severely reduced in its e ciency. owever, the gradually decreasing e ciency of e shell burning causes the envelope to contract, so that the burning shell recovers its role in determining the star's luminosity. Following this contraction, the convective envelope penetrates toward the elium and carbon rich layers. hen this happens, protons are mixed into these layers (mar ed proton mixed downward in Fig. 2) and can interact with carbon to produce ^{14}N by the N cycle, and also neutrons by the reaction the reaction chain shown in Fig.2. The produced neutrons enable the formation of the main component of the s process, and the production of the element ^{19}F as well.

The main problem with this scenario is to get the right amount of ^{13} during the interphase period. It is a problem of the proton mixing into the carbon rich layers.Too many mixed protons is not desirable, since the N cycle would operate producing ^{14}N, which acts as neutron poison. hile few mixed protons is also not desirable, since not enough amount of ^{13} is produced. But how are the protons mixed and how extended is the mixed region . ne way is to use a di usion process for proton mixing with exponentially decreasing di usion coe cient. This treatment has been suggested by the multidimensional fluid simulation of 13 . owever, to use such a description in the one dimensional calculation can only be done in a parameteri ed way.

e cannot discuss all aspects of neutron production in the AGB phases in this short contribution. e ust summari e interesting features of recent investigations. In the wor by

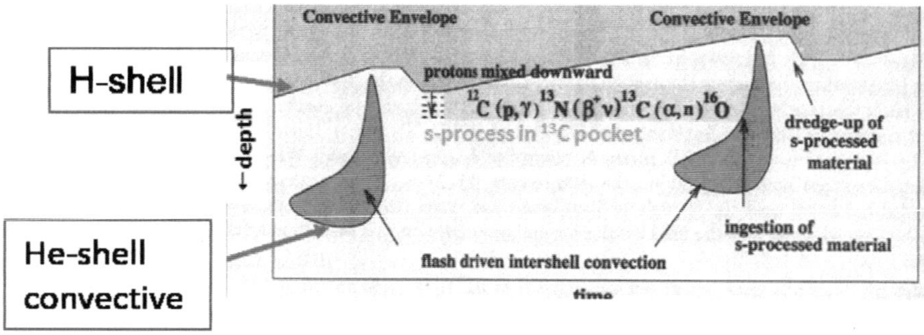

Figure 2. Schematic illustration of tow consecutive thermal pulses of an AGB star. Similar graph is found for example in the paper by 2 , see text for description of the Figure

7 , there is an interesting distinction between the low mass AGB stars (MS) of initial masses $< 4\,M_\odot$ and more massive ones (IMS) with respect to the neutron source. In case of the MS, the neutron source is dominated by the reaction $^{13}C(\alpha,n)^{16}O$, which emphasi es the amount of produced ^{13} , usually termed ^{13} poc et. The lager the core mass on the AGB, the lower the ^{13} amount, owing to the narrower intershell layers. For example, this poc et is four times smaller than in case of a $6\,M_\odot$ star compared to a $2\,M_\odot$ star according to 11 . In contrast, in case of the IMS, larger temperatures are found (7) at the base of the convective e shell during the pulse, so that the reaction $^{22}Ne(\alpha,n)^{25}Mg$ is activated. An indication of this activation is the Rubidium (^{87}Rb) enhancement, whose production is by passed during the radiative ^{13} burning because of the ^{85} r branching (11). It is ama ing to nd out that in case of MS an overshooting at the bottom of the connective e shell could also activate the $^{22}Ne(\alpha,n)^{25}Mg$ reaction (see discussion in 12). Indication of this ind of burning could be done by analy ing the ration of the irconium isotopes $^{96}Zr/^{94}Zr$ which can be measured in Si grains. The point here is that along the r isotopic chain, two unstable isotopes are encountered, ^{93} r with a lifetime of 1.5 My, and ^{95} r with lifetime of 64 day. In other words, only released neutrons from the ^{22}Ne neutron source can produce ^{96} r.

Finally, we mention the most recent considerations, those by and . A new approach is suggested dealing with the ^{13} production. These authors argue that for long time the adopted amount of ^{13} was ta en to be about $1\ ^{-3}\,M_\odot$. owever, an amount about four times larger seems to be re uired to explain the chemical evolution of the s process species.The suggestion is to consider the e ect of the magnetic eld. According to the recent paper by , the buoyancy of magneti ed plasma structures promotes down flow of envelope material to ensure mass conservation. This can lead to a more extended ^{13} poc et. Such consideration rely on the analytic solution of the ideal magnetohydrodynamic e uations presented by 1 . Such calculations as they are presented are free of parametri ation. Multidimensional simulations seem to be useful in order to ustify this conclusion.

4. Summary

Some remar able features of the evolution on the AGB are as follows:

- Stronger e shell flashes give rise to deeper dredge up, which also depends on the envelope mass, thus on the mass loss rate during pulsations.
- ow does a simultaneous treatment of burning and mixing a ect the third dredge up
- higher luminosity during pulsation is related to deeper dredge up
- The neutron source via the $^{13}C(\alpha,n)^{16}O$ reaction is dominant only in the low mass AGB's

of initial mass $< 4\,M_\odot$. For the more massive AGB's , the reaction $^{22}Ne(\alpha,n)^{25}Mg$ becomes an e ective neutron source influencing the branchings along the s process path.

- The e ect of the magnetic eld on the mixing process on the AGB adds a new physical ingredient indeed. As mentioned above, it seems that magnetohydrodynamic processes at the bottom of the convective envelope promote a down flow of the proton rich material into the carbon rich layer to produce larger amount of ^{13} . In this case, no parameteri ed extra mixing at the edge of the envelope would be re uired. A nal word is to say: modeling of the AGB phases os stars remains an interesting and challenging topic in Astrophysics

4.1. Acknowledgments

The author appreciate the travel support of the RB of the American niversity of Beirut, and the organi er of the Santa Tecla meeting for a partial support.

References

1 Rees, M., 2 , Just Six Numbers, Basic Boo s, (New Yor : SA)
2 attan io J. ., ood, .R., 2 4, in Asymptotic Giant BranchStars, eds., abing, . J. and Olofsson ., (Springer:New Yor), p. 23
3 alabi, G. M., l id, M., 2 15, MNRAS, 451, 2 57
4 alabi, G. M., l id, M., hampagne, A. 2 12, ApJ, 761, 1
5 oosley, S. ., eger, A., 2 15, ApJ, 1 , 34
6 hen, M. . etal., 2 14, MNRAS, 44 , 1274
7 ristallo, . et al., 2 15, ApjS, 21 , 4
Tripella, O. etal., 2 14, ApJ, 7 7, 41
Tripella, O. etal., 2 15, ApJ in press
1 Nucci, M. ., Bussu, M., 2 14, ApJ, 7 7, 141
11 Straniero, O. et al., 2 14, ApJ, 7 5, 77
12 erwig, F., 2 5, ARAA, 43, 435
13 Freytag, B., et al., 1 6, A A, 313, 4 7
14 Jones, S., et al., arXiv:151 . 7417v1 astro oh 26Oct 2 15

Journal of Physics: Conference Series **703** (2016) 012006 doi:10.1088/1742-6596/703/1/012006

Nuclear Astrophysics with LUNA

Carlo Broggini
INFN, via Marzolo 8, 35131 Padova, I
E-mail: broggini@pd.infn.it

Abstract. One of the main ingredients of nuclear astrophysics is the knowledge of the thermonuclear reactions which power the stars and synthesize the chemical elements. Deep underground in the Gran Sasso Laboratory the cross section of the key reactions of the proton-proton chain and of the Carbon-Nitrogen-Oxygen (CNO) cycle have been measured right down to the energies of astrophysical interest. The main results obtained during the 'solar' phase of LUNA are reviewed and their influence on our understanding of the properties of the neutrino and of the Sun is discussed. We then describe the current LUNA program mainly devoted to the study of the nucleosynthesis of the light elements in AGB stars and Classical Novae. Finally, the future of LUNA towards the study of helium and carbon burning with a new 3.5 MV accelerator is outlined.

1. Introduction

Only hydrogen, helium and lithium are synthesized in the first minutes after the big-bang. The other elements of the periodic table are produced in the thermonuclear reactions taking place inside the stars. Nuclear astrophysics studies these reactions which provide the power that allows the stars to shine over their lifetimes. In particular, the knowledge of the reaction cross-section at the stellar energies is the heart of nuclear astrophysics. The reaction occurs in the hot plasma of a star, with temperatures in the range of tens to hundreds of millions degrees, inside an energy window, the Gamow peak, which is far below the Coulomb energy arising from the repulsion between nuclei. In this region the cross section is given by:

$$\sigma(E) = \frac{S(E)}{E}\, exp(-2\,\pi\,\eta), \tag{1}$$

where $S(E)$ is the astrophysical factor (which contains the nuclear physics information) and η is given by $2\,\pi\,\eta = 31.29\, Z_1\, Z_2(\mu/E)^{1/2}$. Z_1 and Z_2 are the nuclear charges of the interacting particles, μ is the reduced mass (in units of amu), and E is the center of mass energy (in units of keV).

At low energies the cross sections are extremely small, because of the small probability to go through the Coulomb barrier. Such smallness makes the star life-time of the length we observe, but it also makes impossible the direct measurement in the laboratory. The rate of the reactions, characterized by a typical energy release of a few MeV, is too low, down to a few events per year, in order to stand out from the laboratory background. Instead, the observed energy dependence of the cross-section at high energies is extrapolated to the low energy region, leading to substantial uncertainties. LUNA, Laboratory for Underground Nuclear Astrophysics, started about twenty five years ago to run nuclear physics experiments in an extremely low-background

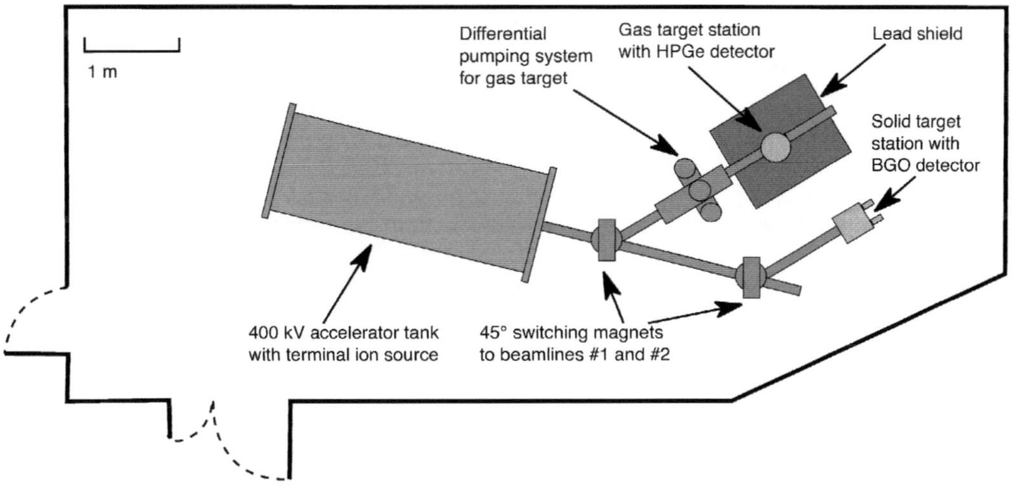

Figure 1. *The LUNA 400 kV accelerator with the two beam lines devoted to gas and solid target experiments, respectively.*

environment, the Gran Sasso Laboratory (LNGS), to reproduce in the laboratory what Nature makes inside the stars [1, 2].

2. LUNA at Gran Sasso

Two electrostatic accelerators able to deliver hydrogen or helium beam have been installed in LUNA: first a compact 50 kV "home made" machine [3] and then a commercial 400 kV one [4]. Common features of the two accelerators are the high beam current, the long term stability and the precise beam energy determination. In particular, the 400 kV accelerator is embedded in a tank, a cylinder of 0.9 m diameter and 2.8 m long, filled with an insulating mixture of N2/CO2 gas at 20 bar. The high voltage is generated by an inline Cockcroft-Walton power supply located inside the tank. The radio frequency ion source directly mounted on the accelerator tube can provide beams of hydrogen and He^+ over a continuous operating time of 40 days. The ions can be sent into one of two different, parallel beam lines (fig.1), allowing the installation of two different target setups. In the energy range between 150 and 400 keV, the accelerator can provide up to 0.5 mA of hydrogen and 0.25 mA of helium at the target stations, with 0.3 keV accuracy on the beam energy, 100 eV energy spread, and 5 eV per hour long-term stability. The dolomite rock of Gran Sasso provides a natural shielding equivalent to at least 3800 meters of water which reduces the muon and neutron fluxes by a factor 10^6 and 10^3, respectively. The activity due to Radon from the rock is suppressed down to the level of few tens of Bequerel/m^3 thanks to frequent air volume exchanges (every 3.5 hours).

3. Hydrogen burning in the Sun

The initial activity of LUNA has been focused on the $^3He(^3He,2p)^4He$ cross section measurement within the solar Gamow peak (15-27 keV). Such a reaction is a key one of the hydrogen burning proton-proton chain (fig.2), which is responsible for more than 99% of the solar luminosity. A resonance in its cross section at the thermal energy of the Sun was suggested long time ago to explain the observed ^8B solar neutrino flux. As a matter of fact, such a resonance would decrease the relative contribution of the alternative reaction $^3He(\alpha,\gamma)^7Be$, which generates the branch responsible for ^7Be and ^8B neutrino production in the Sun.

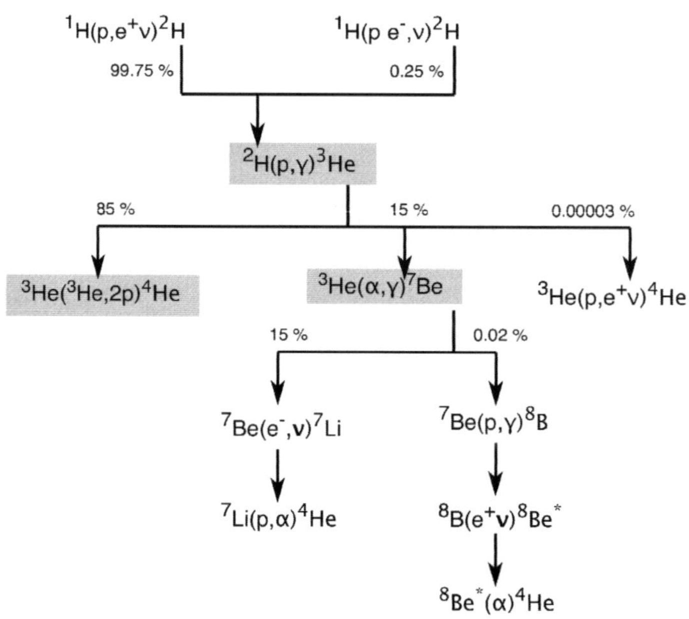

Figure 2. *The proton-proton (pp) chain for hydrogen burning with the different branching ratios. The reactions studied by LUNA are highlighted.*

The experimental set-up was made of eight 1 mm thick silicon detectors of 5x5 cm^2 area placed around the beam inside the windowless target chamber filled with ^{3}He at the pressure of 0.5 mbar. The simultaneous detection of two protons has been the signature which unambiguously identified a ^{3}He(^{3}He,2p)^{4}He fusion reaction. Fig.3 shows the results from LUNA [5] together with higher energy measurements [6, 7, 8]. For the first time a nuclear reaction has been measured in the laboratory at the energy occurring in a star. In particular, at the lowest energy of 16.5 keV the cross section is 0.02 pbarn, which corresponds to a rate of about 2 events/month, rather low even for the "silent" experiments of underground physics. No narrow resonance has been found and, as a consequence, the astrophysical solution of the ^{8}B and ^{7}Be solar neutrino problem based on its existence has been definitely ruled out.

^{3}He(α,γ)^{7}Be, the competing reaction for ^{3}He burning, has also been measured by LUNA both by detecting the prompt γ rays and by counting the ^{7}Be nuclei from their decay. The two different methods gave the same result within the total error of 4% [9].

3.1. The carbon and nitrogen content of the Sun core

^{14}N(p,γ)^{15}O is the slowest reaction of the CNO cycle (fig.4) and it rules its energy production rate. In particular, it is the key reaction to predict the ^{13}N and ^{15}O solar neutrino flux, which depends almost linearly on its cross section.

In the first phase of the LUNA study, data have been obtained down to 119 keV energy with solid targets of TiN and a 126% germanium detector. This way, the five different radiative capture transitions which contribute to the ^{14}N(p,γ)^{15}O cross section at low energy were measured. The total cross section was then studied down to very low energy in the second phase of the experiment by using a 4π BGO summing detector placed around a windowless gas target filled with nitrogen at 1 mbar pressure. At the lowest center of mass energy of 70 keV a cross section of 0.24 pbarn was measured, with an event rate of 11 counts/day from the reaction.

The results obtained first with the germanium detector [10, 11] and then with the BGO set-up

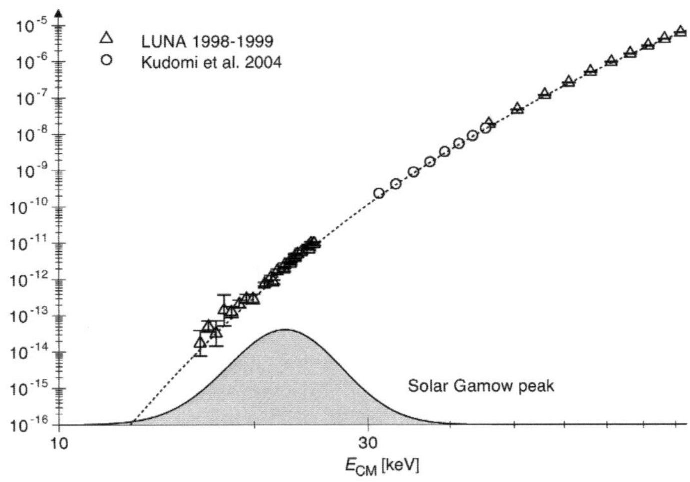

Figure 3. *The cross section of $^3He(^3He,2p)^4He$ as function of energy.*

[12] were about a factor two lower than the existing extrapolation [13, 14] from previous data [15, 16] at very low energy. On the other hand, they were in good agreement with the reanalysis [17] of [16] and with the results obtained with indirect methods [18]. Because of this reduction the CNO neutrino yield in the Sun is decreased by about a factor of two.

In order to provide more precise data for the ground state capture, the most difficult one to be measured because of the summing problem, we performed a third phase of the $^{14}N(p,\gamma)^{15}O$ study with a composite germanium detector. This way the total error on the S-factor has been reduced to 8%: $S_{1,14}(0)=1.57\pm0.13$ keV barn [19]. This is significant because, finally solved the solar neutrino problem, we are now facing the solar composition problem: the conflict between helioseismology and the new metal abundances (i.e. the amount of elements different from hydrogen and helium) that emerged from improved modeling of the photosphere. Thanks to the relatively small error, it will be possible in the near future to measure the carbon and nitrogen content of the Sun core by comparing the predicted CNO neutrino flux with the measured one. As a matter of fact, the CNO neutrino flux is decreased by about 30% in going from the high to the low metallicity scenario. This way it will be possible to test whether the early Sun was chemically homogeneous, a key assumption of the standard Solar Model [20].

The lower cross section is affecting also stars which are more evolved than our Sun. In particular, the lower limit on the age of the Universe inferred from the age of the oldest stellar populations, the globular clusters, is increased by 0.7-1 billion years [21] up to 14 billion years and the dredge-up of carbon to the surface of asymptotic giant branch (AGB) stars is more efficient [22].

4. Hydrogen burning in AGB stars and Classical Novae

A new and rich program of nuclear astrophysics mainly devoted to CNO, Ne-Na and Mg-Al cycles started a few years ago after the solar phase of LUNA. Of particular interest are those bridge reactions which are connecting one cycle to the next, as $^{15}N(p,\gamma)^{16}O$ [23] and $^{17}O(p,\gamma)^{18}F$ [24], the latter competing with $^{17}O(p,\alpha)^{14}N$ [25], or which are key ingredients of gamma astronomy, as $^{25}Mg(p,\gamma)^{26}Al$ [26]. Due to the higher Coulomb barrier of the reactions involved, the cycles become important at temperatures higher than the one of our Sun: hydrogen burning in the shell of massive stars and Novae explosions (about 30-100, and 100-400 million degrees, respectively). Relatively unimportant for energy generation, these cycles are essential for the 'cooking' of the

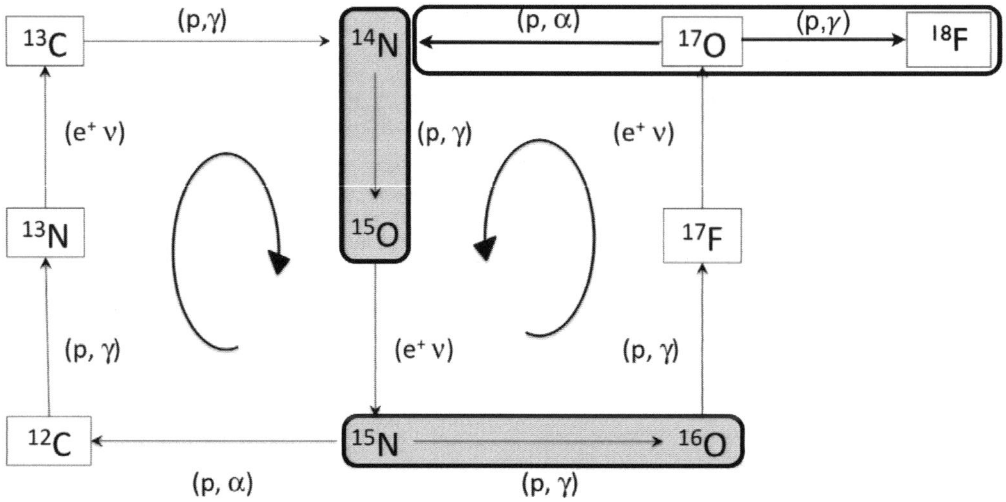

Figure 4. *The first and second CNO cycles. The reactions studied by LUNA are highlighted.*

light nuclei up to ^{27}Al. In particular, LUNA is now measuring ^{22}Ne(p,γ)^{23}Na [27], the reaction of the Ne-Na cycle with the highest uncertainty (up to a factor of 2000 in the region of interest), and ^{23}Na(p,γ)^{24}Mg, the reaction connecting the Ne-Na and Mg-Al cycles. In particular, three low energy resonances of ^{22}Ne(p,γ)^{23}Na have been measured for the first time in the energy region of AGB and Classical Nova burning [28], giving rise to a reaction rate which is a factor of 5 higher than what recently evaluated.

LUNA has achieved significant results also for Big Bang Nucleosynthesis (BBN). In particular, ^{2}H(α,γ)^{6}Li has recently been measured for the first time in the BBN energy region [29]. The results clearly exclude the strong cross section enhancement which has been required as nuclear solution to the primordial ^{6}Li problem (i.e. a ^{6}Li abundance in very old stars which seems to be 3 orders of magnitude higher than predicted).

5. LUNA-MV: the study of helium and carbon burning

After hydrogen burning the natural evolution of LUNA is the study of the next steps in the fusion chain towards ^{56}Fe: helium and carbon burning. In particular, ^{12}C(α,γ)^{16}O determines the abundance ratio between carbon and oxygen, the two key elements to the development of life, and it shapes the nucleosynthesis in massive stars and the properties of supernovae. Equally important are ^{13}C(α,n)^{16}O and ^{22}Ne(α,n)^{25}Mg, the sources of the neutrons which synthesize half of the trans-iron elements through the S-process: neutron capture followed by β decay.

Finally, the ^{12}C+^{12}C fusion reactions are switching on the carbon burning. Their rate determines the evolution of a massive star up to a slowly cooling white dwarf or up to a core-collapse supernova. It also affects the ignition conditions and time scales of thermonuclear supernovae, the standard candles of Cosmology.

This program requires a new 3.5 MV accelerator which is going to be installed underground in hall C of Gran Sasso at the beginning of 2018. The new facility will occupy a surface of about 300 m^2 and it will be contained inside a 5 m high building with thick concrete walls of about 1 m. It will host a single ended electrostatic accelerator able to deliver H, He$^+$, C$^+$ and C^{++} beams at high current with great stability. After a few months of commissioning, we plan to start the first physics runs in summer 2018.

6. Conclusions

Almost twenty five years ago LUNA started underground nuclear astrophysics in the core of Gran Sasso and it still remains the unique facility of this kind in the world. The extremely low background has allowed experiments with count rates as low as two events per month. As a consequence, the important reactions which are responsible for the hydrogen burning in the Sun have been studied for the first time down to the relevant stellar energies. Since a few years LUNA is studying the hydrogen burning reactions which are responsible for the 'cooking' of the light elements in AGB stars and Novae. The future of LUNA, which is going to start with the installation of a 3.5 MV accelerator underground in Gran Sasso at the beginning of 2018, will be the study of helium and carbon burning.

References

[1] H. Costantini *et al*, Rep. Prog. Phys. **72**, 086301 (2009).
[2] C. Broggini *et al*, Annu. Rev. Nucl. Part. Sci. **60**, 53 (2010).
[3] U. Greife Greife *et al*, Nucl. Instrum. Methods A **350**, 327 (1994).
[4] A. Formicola *et al*, (2003), Nucl. Instrum. Methods A **507**, 609 (2003).
[5] R. Bonetti *et al*, Phys. Rev. Lett. **82**, 5205 (1999).
[6] A. Krauss *et al*, Nucl. Phys. A**467**, 273 (1987).
[7] M. Dwarakanath and H. Winkler, Phys. Rev. C **4**, 1532 (1971).
[8] N. Kudomi *et al*, Phys. Rev. C **69**, 015802 (2004).
[9] F. Confortola *et al*, Phys. Rev. C **75**, 065803 (2007).
[10] C. Formicola *et al*, Phys. Lett. B **591**, 61 (2004).
[11] G. Imbriani *et al*, Eur. Phys. J. A **25**, 455 (2005).
[12] A. Lemut *et al*, Phys. Lett. B **634**, 483 (2006).
[13] E.G. Adelberger *et al*, Rev. Mod. Phys. **70**, 1265 (1998).
[14] C. Angulo *et al*, Nucl. Phys. A **656**, 3 (1999).
[15] W.A.S. Lamb and R.E. Hexter, Phys. Rev. **108**, 1304, (1957).
[16] U. Schroder *et al*, Nucl. Phys. A **467**, 240 (1987).
[17] C. Angulo and P. Descouvement, Nucl. Phys. A **690**, 755 (2001).
[18] A.M. Mukhamedzhanov *et al*, Phys. Rev. C **67**, 065804 (2003).
[19] M. Marta *et al*, Phys. Rev. C **78**, 022802 (2008).
[20] W.C. Haxton and A.M. Serenelli, Astrophysical Journal **687**, 678 (2008).
[21] G. Imbriani *et al*, Astronomy and Astrophysics **420**, 625 (2004).
[22] F. Herwig and S.M. Austin, Astrophysical Journal **613**, L73 (2004).
[23] A. Caciolli *et al*, Astronomy and Astrophysics **533**, A66 (2011).
[24] D.A. Scott *et al*, Phys. Rev. Lett. **109**, 202501 (2012).
[25] C. G. Bruno *et al*, Eur. Phys. J. A **51**, 94 (2015).
[26] O. Straniero *et al*, Astrophysical Journal **763**, 100 (2013).
[27] F. Cavanna *et al*, Eur. Phys. J. A **50**, 179 (2014).
[28] F. Cavanna *et al*, to be published in Phys. Rev. Lett., arXiv:1511.05329.
[29] M. Anders *et al*, Phys. Rev. Lett. **113**, 042501 (2014).

Journal of Physics: Conference Series **703** (2016) 012007 doi:10.1088/1742-6596/703/1/012007

Indirect methods in nuclear astrophysics

C.A. Bertulani[(a,b)], Shubhchintak[(a)], A. Mukhamedzhanov[(b)], A. S. Kadyrov[(c)], A. Kruppa[d)], and D. Y. Pang[(e)]

[(a)]Department of Physics and Astronomy, Texas A&M University-Commerce, Commerce, TX 75429, USA
[(b)]Department of Physics and Astronomy, Texas A&M University, College Station, TX 75429, USA
[(c)]Department of Physics and Astronomy, Curtin University, GPO Box U1987, Perth 6845, Australia
[(d)]Institute for Nuclear Research, Hungarian Academy of Sciences, Debrecen, PO Box 51, H-4001, Hungary
[(e)]School of Physics and Nuclear Energy Engineering, Beihang University, Beijing 100191, China

E-mail: `carlos.bertulani@tamuc.edu`, `shub.shubhchintak@tamuc.edu`

Abstract. We discuss recent developments in indirect methods used in nuclear astrophysics to determine the capture cross sections and subsequent rates of various stellar burning processes, when it is difficult to perform the corresponding direct measurements. We discuss in brief, the basic concepts of Asymptotic Normalization Coefficients, the Trojan Horse Method, the Coulomb Dissociation Method, (d,p), and charge-exchange reactions.

1. Introduction

In astrophysics there is a plethora of interesting questions related to the evolution of the Universe, of processes responsible for energy generation in stars and how the nuclei that are observed on Earth are formed. Particularly, one is concerned with various processes in the pp-chain, CNO-cycle, r-process, s-process, etc., through which elements are created in the Universe. The knowledge of cross sections of some specific nuclear reactions such as capture reactions (p,γ), (n,γ), (α,γ) give important information about these processes.

Although direct experiment are preferable, in a large number of cases they are difficult to carry out. This is mainly due to the following reasons:

(i) Experiments have to be performed at stellar energies which are usually very small (of the order of tens or hundreds of keV/u). The cross sections for the reactions involving charged particles are very small (nano barns or pico barns). Often, even after long hours of data collection only a few events are obtained. Background and stability problems have to be taken care of [1]. Direct measurements at low energies can also be affected by electron screening, requiring a difficult and uncertain treatment. An alternative way is to perform experiments at higher energies (few MeV/u) and then extrapolate the results down to the desired energies. But this procedure also involves considerable uncertainties.

(ii) Many of the stellar reactions involve unstable nuclei with very short life time.

2. Indirect methods

To mitigate these problems, alternate indirect methods (a combination of experimental and theoretical analysis) are currently being used where equivalent information can be extracted by performing reactions at much higher beam energies. The indirect methods depend upon the type of reaction involved. Of these, three indirect methods stand out in present day nuclear astrophysics: (a) Asymptotic Normalization Coefficient method, (b) Trojan horse method and (c) Coulomb dissociation method. See, e.g., the review articles [2, 3].

2.1. Asymptotic Normalization Coefficient

The Asymptotic Normalization Coefficient (ANC) method is based on the normalization of the tail of the quantum overlap of bound state wave functions of the initial and final nuclei [2, 4]. Capture reactions in stellar environments take place either by direct capture or by resonant capture. Direct capture reactions of charged particles usually involve systems with small binding energies and they are mostly peripheral at stellar energies. On the other hand, neutron capture reactions may contain contributions from the nuclear interior and are very sensitive to the spectroscopic factor (SF) of the final state [5]. The presence of the Coulomb barrier for charged particle capture causes the reaction to be peripheral. Classically, if the incident particle energy is below the barrier then no capture will occur, however quantum mechanical 'tunneling' gives a probability of barrier penetration. Therefore, the capture in such case proceeds through the tail of the nuclear overlap of the initial and final bound state wave functions whose shape is completely determined by the Coulomb force and the amplitude of the tail is given by the ANC. Although the ANC method has been mostly used for charged particle direct capture reactions, it can also be used for neutron capture (n,γ) reactions, as explained in Ref. [5]. One can extract the ANC from differential cross section data of peripheral particle transfer reactions having the same vertex or from the one-nucleon breakup reaction of loosely bound nuclei.

Consider the virtual decay process $B \to A + a$. If $\phi_B(\zeta_a, \zeta_A; \mathbf{r})$, $\phi_a(\zeta_a)$ and $\phi_A(\zeta_A)$ are the bound state wave functions for respective nuclei with ζ's being the internal coordinates, then the overlap function is given by $I^B_{Aa}(\mathbf{r}) = \langle \phi_A(\zeta_A)\phi_a(\zeta_a)|\phi_B(\zeta_a, \zeta_A; \mathbf{r})\rangle$. In the asymptotic limit (when the nuclear forces are vanishingly small), the radial part $I^B_{Aa}(r)$ of the overlap function can be expressed in terms of Whittaker function ($W_{a,b}$),

$$I^B_{Aa}(r) = C^B_{Aa\ell_B j_B} \frac{W_{-\eta,\ell_B+1/2}(2kr)}{r}; \qquad r > R_N, \tag{1}$$

where R_N is the nuclear interaction radius of A and a, C is the ANC. It defines the amplitude of the tail of the radial overlap integral and it depends upon the structure of nucleus B. k and η are the wave number and the Coulomb parameter for the bound state (Aa), respectively. ℓ_B and j_B are the orbital and total angular momentum of B.

The above formalism of ANC is based on the overlap integral and is also called Schrödinger formalism as it involves wave functions and potentials. However, there is another way to incorporate ANC in the theory of direct reactions. This formalism is based on general scattering theory and it is valid both in non-relativistic quantum mechanics as well as in field theory. Here, the ANC can be used to find the residues at the poles (for bound or resonance states) of the S-matrix for elastic scattering of the $a+A$ system. For more details about this method, see Ref. [2]. The cross section for the direct radiative capture reaction $A+a \to B+\gamma$ can be written as

$$\sigma(E_a) = KF \left| \left\langle I^B_{Aa}(r) \left| \hat{O}(EL) \right| \chi^{(+)}_i(r) \right\rangle \right|^2, \tag{2}$$

where KF is a kinematic factor, $\chi^{(+)}_i$ is the scattering wave function in the initial channel and $\hat{O}(EL)$ is the electromagnetic transition operator connecting the initial and final channel and it does not involve the structure of the core A. In the asymptotic limits the capture cross section [Eq. (2)] can be written as

$$\sigma(E_a) = KF\left|\left\langle C^B_{A a \ell_B j_B} W_{-\eta,\ell_B+1/2}(2kr)/r \left|\hat{O}(EL)\right| \chi^{(+)}(r)\right\rangle\right|^2 = (C^B_{A a \ell_B j_B})^2 w(E_a), \quad (3)$$

where $w(E_a)$ is a function independent over the bound state structure of B. Thus, the direct capture cross section completely depends upon the ANC. By determining the ANC from some other direct reaction experiment (e.g., from elastic scattering), one can calculate the capture cross section at stellar energies. However, one should keep in mind the necessary conditions that the reaction chosen for extracting the ANC should have the same vertex and also that it should be peripheral in nature. The common choices are (i) a transfer reaction between the heavy ions, having one of the vertices kept the same as the vertex for the desired direct capture reaction and (ii) a breakup reaction of loosely bound nuclei.

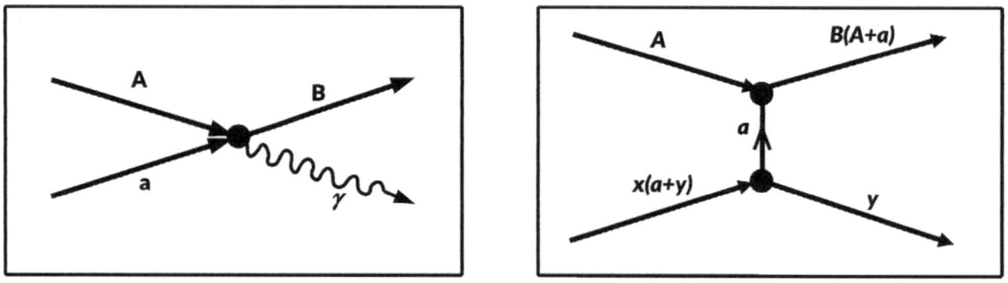

Figure 1. (a) The reaction vertex in the radiative capture reaction $A(a,\gamma)B$ and (b) the reaction vertices in the transfer reaction $A(x,y)B$.

ANC from transfer reactions. Consider the transfer reaction $A(x,y)B$, with $x = y + a$ and $B = A + a$. Figure 1 shows the vertex (vertices) for the direct radiative capture (transfer) reaction. The distorted-wave Born approximation (DWBA) is used to analyze such type of peripheral transfer reactions. The DWBA cross section (σ^{DW}) for the transfer reaction can be related with the measured angular distribution of the nuclei by means of the relation

$$\frac{d\sigma}{d\Omega} = (S^B_{A a \ell_B j_B})(S^x_{y a \ell_x j_x})\sigma^{DW}_{\ell_B j_B \ell_x j_x}, \quad (4)$$

where S^i, ℓ_i and j_i ($i = B$ and x) are the SF, and the orbital and total angular momentum of the respective nuclei. These SF's are related to the corresponding ANC's of the overlap function by

$$(C^B_{A a \ell_B j_B})^2 = (S^B_{A a \ell_B j_B})(b^B_{A a \ell_B j_B})^2; \qquad (C^x_{y a \ell_x j_x})^2 = (S^x_{y a \ell_x j_x})(b^x_{y a \ell_x j_x})^2,$$

where b's are the single particle ANC's of the particle a in the nuclei x and B whereas, C^x and C^B are the ANC's of the initial and final nuclei. Therefore, Eq. (4) can be written in terms of ANC's as

$$\frac{d\sigma}{d\Omega} = (C^B_{A a \ell_B j_B})^2 (C^x_{y a \ell_x j_x})^2 \frac{\sigma^{DW}_{\ell_B j_B \ell_x j_x}}{b^2_{A a j_B \ell_B} b^2_{y a \ell_x j_x}}. \quad (5)$$

When more than one orbital participate, a summation over all the orbitals is required. For peripheral reactions, the ratio in Eq. (5) is independent of the single particle ANC's (b's in the denominator), which also gives a criterion for peripheral reactions. Therefore, by comparing the DWBA cross section with the experimental result one can extract the spectroscopic factor and hence the ANC ($C^{B}_{Aa\ell_{B}j_{B}}$) for the vertex of interest ($A + a \rightarrow B$), provided that the information about the other vertex ($x \rightarrow y + a$) (i.e. $C^{x}_{ya\ell_{x}j_{x}}$) is already known from other independent measurements. This method has been tested in several direct capture reactions to obtain astrophysical S-factors [6].

ANC from the single nucleon removal reactions. It is a well known fact that loosely bound nuclei have one or two nucleon(s) well detached from core nucleons and that they can be easily removed in the nuclear field of a target. Such reactions are usually peripheral even at energies of few hundred MeV/nucleon and they are dominated by the asymptotic tail of the removed nucleon wave function. From the measurement of momentum distributions of the core, one can extract the spectroscopic information about the relative orbital angular momentum of the fragments and contribution of the different single particle states [7]. From the measured one-nucleon removal cross section one can also extract the ANC for the capture process of interest [8]. The one-nucleon removal cross section (σ) for the reaction $B \rightarrow A + a$ can be written as $\sigma = (C^{B}_{Aa\ell j})^{2}\sigma_{sp}/(b^{B}_{Aa\ell j})^{2}$, where σ_{sp} is the calculated single particle removal cross section. One has to take the sum over all the orbitals contributing to the total cross section. The advantage of this technique is that the one-nucleon removal cross sections are usually large even at high energies and therefore a beam of lesser quality can be used.

2.2. Trojan Horse method

This is an another indirect method, where the cross section of a suitable two-body to three-body reaction ($A+x \rightarrow c+B+y$) is used to extract the cross section for a relevant two-body reaction ($A + a \rightarrow c + B$) at stellar energies. This specific type of reaction (two-body to three-body) is called Trojan Horse (TH) reaction and hence the method is known as the Trojan Horse Method (THM) [10]. A TH reaction is chosen in such a way that nucleus x contains a in its cluster structure ($x = y+a$) and this cluster a interacts with A, whereas the other cluster y remains as a spectator. For this to happen, one has to keep Quasi-Free (QF) kinematical conditions for the clusters y and a, which is based on the consideration that the relative momentum between these clusters in the three-body phase space should be zero or small compared to the wave number of bound state (ay) [9]. In that case the interaction between y and a would be small and hence they will stay at maximal distance from each other. In the above reaction if c stands for γ then it represents a radiative capture reaction. In fact, this method is very flexible and can be used to study many other type of reactions and is not only limited to the radiative processes, such as the Coulomb dissociation and the ANC method. While performing an experiment with a TH reaction, one has to pay attention to all the other possible reactions which can be triggered and that can act as background with respect to the desired reaction event.

As mentioned earlier, the direct measurement of a capture cross section for the binary process $A+a \rightarrow c+B$ at stellar energies is difficult because one has to take into consideration the electron screening effect and the extrapolation to low energies is not reliable because of unexpected resonances [3]. However, one can use the THM to obtain cross sections down to the relevant ultra-low stellar energies. This is because the TH reaction between x and A takes place at energies above the Coulomb barrier so that A can dig into x and there is no possibility of additional an Coulomb barrier between A and a. Here, a acts as a virtual particle and hence does not obey the energy-momentum relations for a free particle. Therefore, the cross sections of the binary sub-process extracted from the TH processes do not contain the related Coulomb barrier and are free from electron screening. Consequently, with this method one can determine the astrophysical S-factor S(E) of the binary process $A(a, c)B$ at low relative kinetic energies of the particles a and A without being affected by electron screening. In addition, comparing the direct capture cross section with those obtained from the THM, one can determine the screening

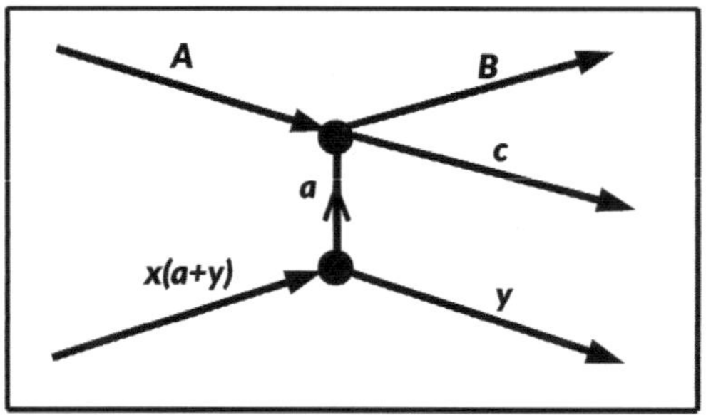

Figure 2. Trojan horse reaction, with three particles in the final state

potential.

As explained in Ref. [12], in spite of the high energy (E_A) of a projectile A (above the Coulomb barrier of $A+x$), the binary reaction $A+a$ still takes place at low relative energies E_{Aa} because of the QF condition. Figure 2 describes the mechanism for the THM. If ϵ_{ay} is the binding energy of intercluster $a-y$ in nucleus x then the relative energy of $A-a$ is given by

$$E_{Aa} = \frac{p_{Aa}^2}{2\mu_{Aa}} - \frac{p_{ay}^2}{2\mu_{ay}} - \epsilon_{ay}, \tag{6}$$

where p's and μ's are the relative momenta and reduced masses of the respective pair. With the QF kinematics, $p_{ay}=0$, and also considering target x at rest (i.e. momentum $P_x=0$) in the lab system, the relative energy E_{Aa} from Eq (6) is

$$E_{Aa} = \frac{m_a}{m_a+m_A}E_A - \epsilon_{ay}, \tag{7}$$

with m's being the masses of respective nuclei. Thus, it is clear that due to the presence of the factor $m_a/(m_a+m_A)$ and the binding energy $-\epsilon_{ay}$, the relative energy E_{Aa} for the binary reaction remains very small and even negative in spite of high E_A.

As mentioned earlier, in the QF kinematics, there is negligible interaction of the spectator y with B and c and hence in the final three-body state only the scattering of two fragments c and B is important. Therefore, the plane-wave-impulse approximation (PWIA) is used to analyze the experimental results in the THM. In this approximation, the three-body triple differential cross section for the TH reaction is factorized as [12, 13]

$$\frac{d^3\sigma}{dE_c d\Omega_c d\Omega_B} = KF|\phi_x(k_{ay})|^2\Big(\frac{d\sigma^{HOES}}{d\Omega_{c.m.}}\Big), \tag{8}$$

where

(i) KF is the kinematical factor which contains the final state phase space factor in terms of masses, angles and momenta of outgoing particles (see Ref. [12]).

(ii) $\phi_x(k_{ay})$ is the Fourier transform of the bound state wave function of x which describes the momentum distribution of x at the final momentum of $a-y$ intercluster. For a nucleus x having ground state angular momentum $\ell=0$, the application of this method is much simpler as the Fourier transform of the ground state wave function peaks at zero momentum and is well known. This helps in determining the QF scattering angles. On the other hand, for high-ℓ nuclei, the momentum distribution becomes much broader and hence the internal separation of the fragments $a-y$ will be small and y can no longer be treated as spectator. Therefore, high-ℓ (>0) nuclei are less suitable for THM (see Ref. [9]).

Journal of Physics: Conference Series **703** (2016) 012007 doi:10.1088/1742-6596/703/1/012007

(iii) $d\sigma^{HOES}/d\Omega$ is the half-off-energy-shell (HOES) differential cross section for the two body reaction $A + a \to B + c$ at relative kinetic energy E_{QF} of the $A - a$ system, such that $E_{QF} = E_{cB} - Q$. Here, E_{cB} is the kinetic energy of the relative motion of c and B and Q is the Q-value of the binary reaction.

It is clear from Eq. (8) that after calculating the factors KF and $|\phi_x(k_{ay})|^2$, the HOES binary cross section can be obtained from the measured triple differential cross section of the TH reaction as

$$\frac{d\sigma^{HOES}}{d\Omega_{c.m.}} \propto \frac{d^3\sigma}{dE_c d\Omega_c d\Omega_B} \frac{1}{KF|\phi_x(k_{ay})|^2}. \tag{9}$$

This binary HOES cross section is then used to extract the quantity of interest, namely, the on-energy-shell cross section by normalization to the direct data available at higher energies. The THM has been used in many theoretical and experimental studies of capture reactions with success. Although, initially it has been used for charged particle capture reactions, now it has also been extended to study direct neutron and resonant capture reactions.

2.3. Coulomb dissociation method

The Coulomb dissociation/breakup method (CD) is an another indirect method proposed by Baur *et al* [14]. As Coulomb induced dissociation is an inverse to the radiative capture process, one can relate the measured Coulomb dissociation cross section (which can be measured at high beam energies and are larger in magnitude) to the relevant radiative capture cross section of astrophysical interest. In this method the composite projectile a consisting of two sub-structures b and c, is broken up in the Coulomb field (or electromagnetic field) of a heavy target nucleus T (see Fig. 3), $a + T \to b + c + T$, where b is the core nucleus and c is a valence nucleon which can be charged or uncharged. The advantage of the CD method is that the measurements are performed at even higher beam energies and the cross sections are considerably larger as the energy increases. Also at sufficiently high incident velocities the fragments b and c emerge with high energies which in turn facilitates their detection. With adequate kinematical conditions in coincidence measurements the study of low relative energies of the final state fragments is still possible [14]. However, for the viability of the method one has to be ensure that the nuclear breakup effect should be negligible or should be known accurately. For a pure Coulomb breakup, a large impact parameter or small scattering angle is often required. Furthermore, the breakup process is considered to be elastic, so that there will be no excitations in the target, although this condition can be relaxed if the energy transferred to the excitation is small compared to the bombarding energy, what is usually the experimental choice.

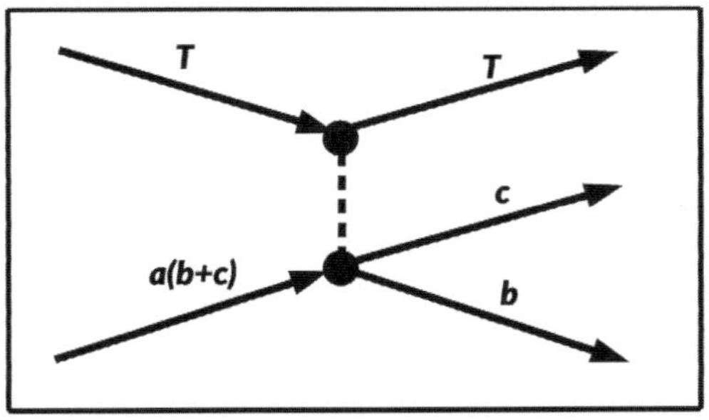

Figure 3. Coulomb breakup of nucleus a in the electromagnetic field of heavy target T.

The Coulomb breakup process [Eq. (2.3)], is then related to the photodisintegration reaction $a+\gamma \rightarrow b+c$, where one calculate the photodisintegration cross sections ($\sigma^{\pi\lambda}_{\gamma,n}$) as

$$\frac{d\sigma}{dE_{rel}} = \frac{1}{E_\gamma}\sum_{\pi\lambda}\sigma^{\pi\lambda}_{\gamma,n}\,\mathrm{n}_{\pi\lambda}, \tag{10}$$

where $d\sigma/dE_{rel}$ is the relative energy spectra in the Coulomb breakup at relative energy E_{rel} of $b-c$ system, π stands for either electric or magnetic transition of multipolarity λ, $E_\gamma = E_{rel}+Q$ is the photon energy with Q as the Q-value of the reaction and $\mathrm{n}_{\pi\lambda}$ is the equivalent photon number (also called virtual photon number) and depends on the $a-t$ system (see Ref. [15]). Similar to Eq. (10) the angular differential cross section of the Coulomb breakup can also be related to photo disintegration cross section as

$$\frac{d\sigma}{d\Omega} = \frac{1}{E_\gamma}\sum_{\pi\lambda}\sigma^{\pi\lambda}_{\gamma,n}\frac{d\mathrm{n}_{\pi\lambda}}{d\Omega}, \tag{11}$$

Ω being the scattering solid angle.

It has been found that relativistic effects, higher order and higher multipole's contribution can influence the reaction process [16, 17] and hence make it difficult to use Eqs. (10)-(11). One has to take these effects into account while using the CD method. When a single multipolarity of either type (electric or magnetic) dominates, the above equations can be easily used to find the photodisintegration cross section for the dominating multipolarity (λ). Using the 'principle of detailed balance' the capture cross section ($\sigma_{n,\gamma}$) for the desired capture process $b+c \rightarrow a+\gamma$, can be obtained from $\sigma^{\pi\lambda}_{\gamma,n}$ as

$$\sigma_{n,\gamma} = \frac{2(2j_a+1)}{(2j_b+1)(2j_c+1)}\frac{k_\gamma^2}{k_{bc}^2}\sigma^{\pi\lambda}_{\gamma,n}, \tag{12}$$

where j_a, j_b and j_c are the spins of particles a, b and c, respectively. k_γ and k_{bc} are the wave numbers of the photon and that of relative motion between b and c, respectively. As a word of caution, the principle of detailed balance cannot be used for the situations where capture leads to the bound excited states having unknown branching ratios of γ-ray emission [18].

The CD method has been used to determine the capture cross sections for numerous radiative capture stellar processes using both experimental and advanced theoretical studies (see Ref. [2]). In fact, there are different theoretical models used to study the Coulomb dissociation. In Refs. [19, 20, 21], a pure quantum mechanical model [23] using the post-form DWBA is used to calculate the radiative neutron capture cross section on ^{8}Li, ^{14}C and ^{15}N from the Coulomb breakup of ^{9}Li, ^{15}C and ^{16}N, respectively. For ^{15}N$(n,\gamma)^{16}$N, there are large contributions from the capture to the low-lying bound excited states of ^{16}N. It is pointed out in Ref. [22] that although theoretically such a study is possible, experimentally it is difficult as it requires additional information on the γ-ray branching ratios for the excited states, similar to what was done in Ref. [24]. Recently, in Refs. [25, 26], the theory of Coulomb dissociation [23] (for neutron exotic nuclei) has been extended to include the projectile deformation. This further opens the door to study the affect of deformation on the neutron capture cross sections and subsequent rates for medium mass nuclei.

3. (d,p) and charge-exchange reactions

In surrogate reactions such as (d,p) reactions, the neutron is brought to react with a nucleus inside a Trojan, or surrogate nucleus (here, the deuteron). Such reactions are thought to give information on neutron-induced reactions, often not available in direct measurements. The neutron inside the deuteron has angular momentum and other quantum numbers that are not equal to those of free neutrons at the same energy. If angular momentum is not relevant in the reaction of interest (a rare situation), the desired neutron-induced reaction can be extracted from the surrogate equivalent. (d,p) reactions are not as simple as one might initially think because they involve at least three particles in the incoming and outgoing channels. Assuming it can be treated as a three-body reaction, one may use the *Alt-Grassberger-Sandhas* (AGS) [27] formalism to describe the reaction. The AGS equations are a branch of the Faddeev formalism to treat the $(a+b)+A$ system, and its rearrangements $a+(b+A)$ and $b+(a+A)$. The three-particle scattering in each channel is obtained from the transition operators $T_{\beta\alpha}$, where $\alpha(\beta)$ corresponds to channel permutation combinations. They are the set of coupled integral equations

$$T_{\beta\alpha}(z) = \tilde{\delta}_{\alpha\beta} G_0^{-1}(z) + \sum_{\nu=1}^{3} \tilde{\delta}_{\beta\nu} t_\nu(z) G_0(z) T_{\nu\alpha}(z) \ , \tag{13}$$

where $\tilde{\delta}_{\beta\nu} = 1-\delta_{\beta\nu}$ is the anti-delta-Kronecker symbol, and $G_0 = (E+i_{\epsilon^+}-H_0)^{-1}$ for the three-free particle c.m. energy E and Hamiltonian H_0. The two-body transition operator t_ν for each interacting pair with inter-potential v_ν is determined from the Lippmann-Schwinger equation $t_\nu = v_\nu + v_\nu G_0 t_\nu$. Additional approximations are often necessary and the correct treatment of the Coulomb interaction poses a major difficulty to solve the AGS set of equations [28]. One can treat the Coulomb interaction with help of screened Coulomb potentials, but this method works only for light nuclei since wild oscillations in the asymptotic region jeopardizes the accuracy of the solutions [29].

In charge-exchange reactions, such as (p,n) reactions, one seeks information on Gamow-Teller matrix elements which cannot be studied in β-decay experiments. The method is useful to obtain cross sections for inelastic neutrino scattering and electron capture reaction rates in stars. Not only (p,n), but (^{3}He,t) and other charge exchange reactions at energies around 100 MeV/nucleon have been used [30]. Experimentally, one often uses simplified theories in the experimental analysis, e.g., the cross section $\sigma(p,n)$ at small momentum transfer q is assumed to be [31],

$$\frac{d\sigma}{dq}(q=0) = KF.N_D |J_{\sigma\tau}|^2 B(\alpha), \tag{14}$$

where KF (N) is a kinematical (distortion) factor, approximately describing initial and final state interactions, $J_{\sigma\tau}$ is the Fourier transform of the charge-exchange interaction, and $B(\alpha = F, GT)$ is the reduced transition probability for non-spin-flip, $B(F) = (2J_i+1)^{-1}|\langle f || \sum_k \tau_k^{(\pm)} || i\rangle|^2$, and spin-flip, $B(GT) = (2J_i+1)^{-1}|\langle f || \sum_k \sigma_k \tau_k^{(\pm)} || i\rangle|^2$, transitions. The simplified approach based on Eq. (14) has been criticized in other theoretical treatments [32] and is currently under further theoretical study [33].

Acknowledgments

C.A.B. and A.M.M. acknowledge funding support by the NSF, PHY-1415656, and by DOE, DE-FG02-08ER41533, DE-FG02-93ER40773 and DE-SC0004958. A.S.K. acknowledges the support by the Australian Research Council. A.K. acknowledges funding support from Hungarian Scientific Research Fund OTKA K112962 and D.Y.P. acknowledges support of the National Natural Science Foundation of China (Grants 11275018 and 11035001) and the Chinese Scholarship Council (Grant 201303070253).

References

[1] Aliotta M, Junker M, Prati P, and Strieder F 2016 Helium burning and neutron sources in the stars *European Physical Journal* - Topical Issue - to be published.

[2] Tribble R E, Bertulani C A, La Cognata M, Mukhamedzhanov A M and Spitaleri C 2014 Indirect techniques in nuclear astrophysics: a review *Rep. Prog. Phys.* **77** 106901-1-49.

[3] Bertulani C A and Gade A 2010 Nuclear astrophysics with radioactive beams *Physics Reports* **485** 195.
[4] Xu H M, Gagliardi C A, Tribble R E, Mukhamedzhanov A M and Timofeyuk N K 1994 Overall Normalization of the Astrophysical S Factor and the Nuclear Vertex Constant for ^{7}Be$(p,\gamma)^8$B Reactions *Phys. Rev. Lett.* **73**, 2027-2030.
[5] Mukhamedzhanov A M, Nunes F M and Mohr P 2008 Benchmark on neutron capture extracted from (d,p) reactions, *Phys. Rev. C* **77** 051601- 1-5 (R).
[6] Mukhamedzhanov A M and Pang D Y 2015 Asymptotic normalization coefficients and radiative widths *Phys. Rev. C* **92** 014625.
[7] Bertulani C A and Hansen P G 2004 Momentum distributions in stripping reactions of radioactive projectiles at intermediate energies *Phys. Rev. C* **70** 034609.
[8] Trache L, Carstoiu , Mukhamedzhanov A M and Tribble R E 2002 Determination of the S_{18} astrophysical factor for ^{8}B$(p,\gamma)^9$C from the breakup of ^{9}C at intermediate energies *Phys. Rev. C* **66** 035801-1-10.
[9] Spitaleri C *et al.* 2000 The $\alpha-^{12}$C scattering studied via the Trojan-Horse method *Eur. Phys. J. A* **7** 181-187.
[10] Baur G 1986 Breakup reactions as an indirect method to investigate low-energy charged-particle reactions relevant for nuclear astrophysics *Phys. Lett. B* **178** 135-138.
[11] Typel S and Baur G 2003 Theory of the Trojan-Horse method *Ann. Phys.* **305** 228-265.
[12] Tumino A *et al.* 2013 New advances in the Trojan Horse Method as an indirect approach to nuclear astrophysics, *Few-Body Syst.* **54** 745-753.
[13] Pizzone R G *et al.* 2014 Trojan horse particle invariance: An extensive study *Few-Body Syst.* **55** 1001-1004.
[14] Baur G, Bertulani C A and Rebel H 1986 Coulomb dissociation as a source of information on radiative capture processes of astrophysical interest *Nucl. Phys. A* **458** 188-204.
[15] Bertulani C A, Baur G 1988 Electromagnetic processes in relativistic heavy ion collisions *Phys. Rep.* **163**, 299-408.
[16] Bertulani C A 2005 Relativistic continuum-continuum coupling in the dissociation of Halo Nuclei *Phys. Rev. Lett.* **94** 072701-1-4.
[17] Bertulani C A 1996 ^{7}Be$(p,\gamma)^8$B cross section from indirect breakup experiments *Z. Phys. A* **356** 293-297.
[18] Motobayashi T 1998 Coulomb dissociation of unstable nuclei - simulation for radiative captures of astrophysical interest *Nucl. Phys. A* **630** 328-339.
[19] Banerjee P, Chatterjee R, Shyam R 2008 Coulomb dissociation of ^{9}Li and the rate of the ^{8}Li$(n,\gamma)^9$Li reaction *Phys. Rev. C* **78** 035804-1-6.
[20] Shubhchintak, Neelam and Chatterjee R 2014 Capture cross section and rate of the ^{14}C$(n,\gamma)^{15}$C reaction from the Coulomb dissociation of ^{15}C *Pramana J. Phys.* **83** 533-543.
[21] Neelam, Shubhchintak and Chatterjee R 2015 Structure effects in the ^{15}N$(n,\gamma)^{16}$N radiative capture reaction from the Coulomb dissociation of ^{16}N *Phys. Rev. C* **92** 044615-1-7.
[22] P. Mohr 2015 Comment on "Structure effects in the ^{15}N$(n,\gamma)^{16}$N radiative capture reaction from the Coulomb dissociation of ^{16}N" *Phys. Rev. C* (under review).
[23] Chatterjee R, Banerjee P, Shyam R 2000 Projectile structure effects in the Coulomb breakup of one-neutron halo nuclei *Nucl. Phys. A* **675** 477-502.
[24] Izsák R *et al.* 2013 Determining the ^{7}Li(n,γ) cross section via Coulomb dissociation of ^{8}Li *Phys. Rev. C* **88** 065808-1-8.
[25] Shubhchintak and Chatterjee R 2014 Deformation effects in the Coulomb breakup of ^{31}Ne *Nucl. Phys. A* **922** 99-111.
[26] Shubhchintak, Neelam, Chatterjee R, Shyam R and Tsushima K 2015 Coulomb breakup of ^{37}Mg and its ground state structure *Nucl. Phys. A* **939** 101-120.
[27] Grassberger P and Sandhas W 1967 Systematical treatment of the non-relativistic n-particle scattering problem *Nucl. Phys. B* **2** 181; Alt E O, Grassberger P, and Sandhas W 1970 *Phys. Rev. C* **1** 85.
[28] Mukhamedzhanov A M, Pang D Y, Bertulani C A, and Kadyrov A S 2014 Surface-integral formalism of deuteron stripping *Phys. Rev. C* **90** 034604.
[29] Deltuva A and Fonseca A C 2009 Three-body Faddeev-Alt-Grassberger-Sandhas approach to direct nuclear reactions *Phys. Rev. C* **79** 014606.
[30] Sasano M *et. al.* 2012 Extraction of Gamow-Teller strength distributions from ^{56}Ni and ^{55}Co via the (p,n) reaction in inverse kinematics *Phys. Rev. C* **86** 015809.
[31] Taddeucci T N, et al. 1987 Energy dependence of the ratio of isovector effective interaction strengths from 0° (p,n) cross sections *Nucl. Phys. A* **469** 125.
[32] Bertulani C A 1993 Charge-exchange reactions in righ-energy heavy ion collisions, *Nucl. Phys. A* **554** 493; Bertulani C A and Lotti P 1997 *Phys. Lett. B* **402** 237; Bertulani C A and Dolci D 2000 *Nucl. Phys. A* **674** 527.
[33] Bertulani C A, Pang D Y, and Tanihata I 2016 *work in progress.*

The role of neutrino-nucleus reactions in supernova dynamics and nucleosynthesis

Karlheinz Langanke and Gabriel Martínez-Pinedo

GSI Helmholtzzentrum für Schwerionenforschung, Darmstadt, Germany, and
Institut für Theoretische Physik, Technische Universität Darmstadt, Darmstadt, Germany

E-mail: k.langanke@gsi.de

Abstract. Neutrino reactions on nuclei play important roles for the dynamics of supernovae and their associated nucleosynthesis. This manuscript summarizes the current status in deriving the relevant cross sections for supernova neutrinos and briefly discusses a few recent advances where

1. Introduction

Neutrinos are key-players for the supernova dynamics [1, 2]. During the collapse phase the main neutrino source is electron capture on nuclei [3, 4]. By lowering the electron-to-nucleon ratio Y_e this process reduces the pressure which electrons can stem against the gravitational collapse of the core. Furthermore, at sufficiently low densities the electron neutrinos generated by the capture process can leave the star unhindered keeping the core at relatively low entropies so that heavy nuclei survive the collapse. At densities in excess of about 10^{12} g/cm^3 neutrinos get trapped in the core, mainly by elastic scattering on nuclei. The thermalization of neutrinos with the other core matter occurs by energy exchange via inelastic scattering on electrons and, in a lesser extent, on nuclei. In the final collapse phase at high densities pair production of neutrinos of all flavors becomes relevant. This occurs mainly by nucleon-nucleon bremsstrahlung [5, 6], but also nuclear deexcitation has been identified recently as an important additional source of neutrinos other than electron neutrinos [7]. After core bounce, energy transport by neutrinos from hotter core regions to the matter behind the stalled shock helps to revive the shock [8, 2]. The dominating processes are absorption of electron neutrinos and anti-neutrinos on neutrons and protons. The competition of these two absorption processes also determines the proton-to-neutron ratio for the subsequent explosive nucleosynthesis processes which might occur either in proton-rich environment (νp process [9, 10, 11]) or in neutron-rich environment. The later scenario has been favored for many years as the possible site for the astrophysical r-process (neutrino-driven wind model, [12]), but recent supernova simulations indicate that the conditions in this so-called neutrino-driven wind are probably only sufficient to support a weak r-process which contributes to the observed r-process abundances up to the barium mass region (second r-process peak [13]). There have been several suggestions how neutrino reactions on nuclei might contribute to supernova r-process nucleosynthesis (e.g. [14, 15, 16]), but recent studies point to no significant influence of these processes. Neutrino-induced spallation reactions, however, are crucial, for the production of selected nuclei (ν process [17]). Finally, the observation of supernova neutrinos by earthbound detectors is an eminent tool to verify our understanding of the supernova dynamics and mechanism. One requisite here is the knowledge of the neutrino reaction cross sections for the nuclei comprising the detector material.

There has been a recent extensive review of neutrino-nucleus reactions and their role in supernovae which might be consulted for more details [18].

2. Cross section models

Supernova neutrinos of interest here have relatively low energies (up to a few 10's of MeV). At these energies the cross sections are dominated by allowed transitions. Forbidden transitions become relevant at the higher neutrino energies and in cases where allowed transitions are strongly suppressed [19]. The Fermi contribution to the cross sections are defined by the position of the Isobaric Analog State and the respective sum rule. Charge-exchange experiments have progressed our understanding of Gamow-Teller (GT) distributions significantly in the last two decades [20]. The distributions are strongly fragmented. This is caused by nucleon-nucleon correlations and is well described by nuclear models like the diagonalization shell model which accounts for such correlations [21, 4, 22]. In fact, the combined progress due to experimental GT data from charge-exchange experiments and their detailed description by shell model calculations (except for a constant renormalization factor) have led to a rather reliable description of stellar electron capture [23, 24, 25, 26] which is the dominating weak interaction process during the collapse phase [3, 2]. For heavier nuclei, the calculations require model spaces which span more than one oscillator shell. Such studies can be performed within the Shell Model Monte Carlo approach [27, 28, 29, 30] or, for selected nuclei, also within the diagonalization shell model (e.g. [31]). The absorption of neutrinos on nuclei is the inverse process of (electron and positron) capture. Its calculation has also benefitted from the advances in describing GT distributions. In supernova simulations it is incorporated via detailed balance with the inverse capture processes. Forbidden transitions become relevant at high enough neutrino energies that reliable cross section calculations require the reproduction of the energy centroids and total strengths of the respective transitions distributions, but not their detailed reproduction. These requirements are fulfilled by the Random Phase Approximation (RPA). Hence a 'hybrid model' has been proposed in which the allowed contributions to the neutrino-nucleus cross sections are calculated by the shell model and the forbidden contributions within the RPA formalism [32, 19].

Inelastic neutrino-nucleus scattering can be also evaluated within the hybrid model ansatz for temperature $T = 0$. Validation for this procedure here can be derived from precision M1 data for spherical nuclei, measured by inelastic electron scattering, which are dominated by the the same nuclear transitions [33]. At stellar temperatures, however, transitions mediated from thermally populated excited nuclear states modify the cross sections at low neutrino energies significantly. Two approaches have been proposed to incorporate these modifications: i) by including selected GT transitions involving excited states [33] and ii) within the consistent extension of the RPA to finite temperatures (Thermal Quasiparticle RPA) [34, 35].

Neutrino-nucleus reactions often excite the daughter nucleus to states above particle thresholds which then subsequently decay by particle emissions. The probabilities for decay into different particle channels can be calculate within the statistical model. Nuclear spallation reactions are important for supernova nucleosynthesis and potentially also as detection signal for certain supernova neutrino detectors.

3. Neutrino-nucleus reactions in supernova dynamic and nucleosynthesis

In this section we briefly summarize selected recent examples in which the role of neutrino-nucleus reactions have been investigated for the dynamics of core-collapse supernovae, for the production of selected nuclei in the ν nucleosynthesis process and for the spectrum of supernovae neutrinos and their observation by earthbound detectors.

Journal of Physics: Conference Series **703** (2016) 012008 doi:10.1088/1742-6596/703/1/012008

3.1. Inelastic neutrino-nucleus reactions

When, with progressing collapse, matter gets denser, neutrino interactions with matter become increasingly important and require careful treatment which is usually done within Boltzmann transport theory [36, 37]. Here elastic neutrino-nucleus and inelastic neutrino-electron scattering have been identified as the sources for neutrino trapping (which occurs at densities around a few 10^{11} g/cm^3) and neutrino thermalization. Inelastic neutrino-nucleus scattering is another mode of energy exchange between neutrinos and matter and its influence on the supernova dynamics has been studied in Ref. [38]. As has been discussed above, the relevant cross sections can be calculated based on large-scale shell model calculations for the allowed GT transitions and within the random phase approximation for forbidden transitions [39].

Although inelastic neutrino-nucleus scattering contributes to the thermalization of neutrinos with the core matter, the inclusion of this process has no significant effect on the collapse trajectories. However, it increases noticeably the opacity for high-energy neutrinos in the ν_e neutrino burst just after the bounce and has important consequences for the resulting supernova neutrino spectrum (see below).

3.2. Neutrino nucleosynthesis

When neutrinos, produced in the hot supernova core, pass through the outer shells of the star, they can induce nuclear reactions and in this way contribute to the elementsynthesis (the ν-process, [17]). For example, the nuclides ^{11}B and ^{19}F are produced by $(\nu, \nu' n)$ and $(\nu, \nu' p)$ reactions on the quite abundant nuclei ^{12}C and ^{20}Ne. These reactions are dominantly induced by ν_μ and ν_τ neutrinos and their antiparticles (combined called ν_x neutrinos) which have larger average energies than ν_e and $\bar{\nu_e}$ neutrinos. As found in detailed stellar evolution studies [40] the rare odd-odd nuclides ^{138}La and ^{180}Ta are mainly made by the charged-current reaction ^{138}Ba$(\nu_e, e^-)^{138}$La and ^{180}Hf$(\nu_e, e^-)^{180}$Ta. Hence, the ν-process is potentially sensitive to the spectra and luminosity of ν_e and ν_x neutrinos, which are the neutrino types not observed from SN1987a.

Recent supernova simulations, with improved descriptions of neutrino matter interactions, indicate that the average neutrino energies are smaller than previously assumed. This should result in reduced neutrino-induced cross sections and hence lower elemental production rates. This has been the motivation of recent neutrino nucleosynthesis studies performed for stars with masses between 15 and 40 $M_\odot$ and including neutrino-nucleus cross sections for a large set of nuclei with $Z < 78$. As an additional improvement in comparison to previous calculations these nucleosynthesis studies considered differential cross sections for multi-particle emissions [41]. Mainly due to the change in neutrino spectra, this study finds slightly smaller abundances for ^{7}Li, ^{11}B, ^{138}La and ^{180}Ta, however, it confirms the production of these nuclides by neutrino nucleosynthesis [42] As is shown in Fig. 1, the study also finds that neutrino-induced reactions, either directly or indirectly by providing an enhanced abundance of light particles, noticeably contribute to the production of the radioactive nuclides ^{22}Na and ^{26}Al, which are both prime candidates for gamma-ray astronomy. However, the studies do not find significant production of two other candidates, ^{44}Ti and ^{60}Fe, due to neutrino-induced reactions.

As a major improvement it has been possible recently to measure the GT$_-$ strengths on ^{138}Ba and ^{180}Hf below the particle thresholds and to convert these data into the relevant (ν_e, e^-) cross sections [43]. It is found that the new cross sections are slightly larger than the RPA predictions.

Due to the expected hierarchy of average energies for supernova neutrinos $(\langle E_{\nu_e} \rangle < \langle E_{\nu_x} \rangle$, neutrino oscillations are expected to increase the average ν_e energy and consequently also the charged-current cross section induced by supernova neutrinos. As pointed out by Kajino and collaborators, this makes the ratio of ^{7}Li and ^{11}B sensitive to the θ_{13} mixing angle and to the mass hierarchy [44, 45, 46, 47]. Despite this intriguing sensitivity, an accurate derivation of the ^{7}Li/^{11}B abundance ratio requires reliable stellar model calculations and neutrino and nuclear

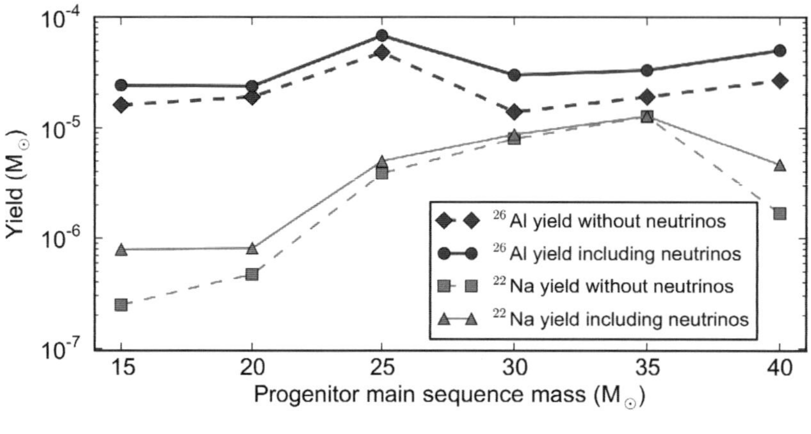

Figure 1. Yields of ^{26}Al (thick lines with circles) and ^{22}Na (thin lines with triangles) for the set of progenitor stars considered. Given is the yield without neutrinos (dashed lines) and including neutrinos with $T_{\nu_e} = 2.8$ MeV and $T_{\bar{\nu}_e} = T_{\nu_\mu,\nu_\tau} = 4MeV$. (from [42])

cross sections, but must also consider the production of the elements from other astrophysical sources; ^{7}Li is, for example, also produced by Big Bang nucleosynthesis [48].

3.3. Detecting supernova neutrinos

The observation of neutrinos from supernova SN1987A by the earthbound detectors Kamiokande [49] and IMB [50] has confirmed and advanced the understanding of core-collapse supernovae. A similar boost is expected from the observation of the next near-by supernova which is likely to test the predictions of supernova models concerning the neutrino spectra for the different flavors, including the noticeable neutrinoburst signal in electron neutrinos [2], originating from electron capture on protons set free by the shock. This spectrum has changed recently after inelastic neutrino-nucleus scattering has been included in supernova simulations [38]. These burst neutrinos traverse regions outside the shock where nuclei have yet not been dissociated and, in particular, high-energy neutrinos excite these nuclei. This means that they are down-scattered in energy, in this way significantly reducing the high-energy tail of the spectrum of emitted supernova neutrinos (see Fig. 2, [38]). This makes the detection of neutrinos, originating from the burst, by earthbound detectors more difficult, as the neutrino detection cross section scales with E_ν^2. Depending on the neutrino reaction threshold the reduction of the expected event rate can be significantly, reaching up to more than 50% for ^{16}O which has been proposed, via nucleon emission following neutral current excitation, as a potential detection scheme for supernova ν_x neutrinos in Superkamiokande [51].

For a recent overview on the detectors and their scheme for observing supernova neutrinos the reader is refered to Ref. [52]. Table 2 of this reference lists the present and future supernova neutrino detectors. Their main material are liquid scintillator (C_nH_{2n}), water, lead or liquid argon. Hence translating the event rates of supernova neutrinos observed in the detectors requires a detailed knowledge of the cross sections for the neutrino interaction with the detector materials.

For ^{12}C shell model calculations have been performed for charged- and neutral-current neutrino reactions in a $(0{+}2)\hbar\omega$ model space [53], improving earlier calculations based on the RPA or on more restricted shell model spaces (see [19]). For the double-magic nucleus ^{16}O

Journal of Physics: Conference Series **703** (2016) 012008 doi:10.1088/1742-6596/703/1/012008

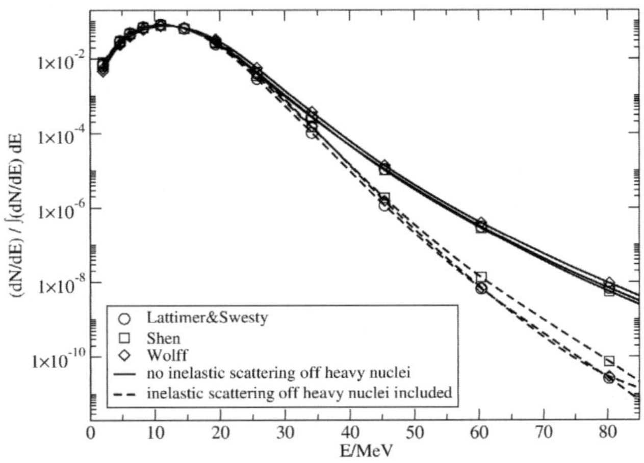

Figure 2. Normalized ν_e number spectra radiated during the shock breakout burst as seen by a distant observer at rest. Results are shown for simulations with three different nuclear EoSs. For better comparison of the strongly time-dependent spectra during this evolution phase, integration in a window of 8 ms around the peak luminosity was performed. Inelastic neutrino scattering off nuclei (dashed lines) leads mostly to energy losses of high-energy neutrinos and thus reduces the high-energy tails of the spectra. The vertical line marks the mean spectral energy. (from [38])

GT transitions are strongly suppressed and the cross sections are dominated by spin-dipole transitions which have recently been modelled in a shell model calculation [54] considering the p and sd shells. Based on a hybrid model approach calculating the GT contribution within the shell model, the transition to the IAS from the Fermi sum rule and the forbidden transitions within the RPA, Suzuki and Honma have determined the ^{40}Ar$(\nu_e, e^-)^{40}$K cross section for neutrino energies up to $E_\nu = 100$ MeV [55]. A similar study of the $(\bar{\nu}_e, e^+)$ cross section on ^{40}Ar has yet not been performed. As the GT contribution to the cross section is strongly suppressed such a study likely requires a shell model calculation performed in the (sd)-(pf) model space for the forbidden transitions. For ^{208}Pb Suzuki and Sagawa presented (ν_e, e^-) cross sections which have been obtained using GT data from a (p, n) experiment and adjusting their Hartree-Fock + Tamm-Dancoff approach for the first-forbidden response to the peaks of the spin-dipole resonances [56]. Furthermore, the spreading and quenching of the GT response has been considered by coupling to 2p-2h configurations. However, the calculations have been performed assuming a muon-decay-at-rest rather than a supernova neutrino spectrum, and are thus not directly relevant to supernova neutrinos.

Acknowledgments
Some of the work reported here has been performed in collaboration with Tobias Fischer, Hans-Thomas Janka and Andre Sieverding whom we gratefully thank. This work was partly supported by the Deutsche Forschungsgemeinschaft through contract SFB 634, the Helmholtz International Center for FAIR within the framework of the LOEWE program launched by the state of Hesse, and the Helmholtz Association through the Nuclear Astrophysics Virtual Institute (VH-VI-417).

[1] H.A. Bethe, Rev. Mod. Phys. **62** (1990) 801
[2] H.-T. Janka, K. Langanke, A. Marek, G. Martínez-Pinedo, B. Müller, Phys. Repts. **442** (2007) 38.
[3] H.A. Bethe, G.E. Brown, J. Applegate and J.M. Lattimer, Nucl. Phys. **A324** (1979) 487
[4] K. Langanke and G. Martinez-Pinedo, Rev. Mod. Phys. **75** (2003) 819
[5] S. Hannestad and G. Raffelt, Astrophys. J. **507** (1998) 339
[6] S. Bacca, K. Hally, M. Liebendörfer, A. Perego, C.J. Pethick and A. Schwenk, Astrophys. J. **758** (2012) 34
[7] T. Fischer, K. Langanke and G. Martinez-Pinedo, Phys. Rev. C **88** (2014) 065804
[8] H. A. Bethe and J. R. Wilson, Astrophys. J. **295** (1985) 14.
[9] C. Fröhlich *et al.*, Astrophys. J. **637** (2005) 415
[10] C. Fröhlich, G. Martinez-Pinedo *et al.*, Phys. Rev. Lett. **96** (2006) 142502
[11] J. Pruet *et al.*, Astrophys. J. **623** (2005) 325
[12] S.E. Woosley *et al.*, Astrophys. J. **433** (1994) 229
[13] C.J. Hansen, F. Montes and A. Arcones, Astr. J. **797** (2014) 123
[14] Y.-Z. Qian *et al.*, Phys. Rev. **C55** (1997) 1532
[15] W.C. Haxton *et al.*, Phys. Rev. Lett. **78** (1997) 2694
[16] M. Terasawa *et al.*, Astrophys. J. **608** (2004) 470
[17] S.E. Woosley *et al.*, Astrophys. J. **356** (1990) 272
[18] K.G. Balasi, K. Langanke and G. Martinez-Pinedo, Prog. Part. Nucl. Phys. **85** (2015) 33
[19] E. Kolbe, K. Langanke, G. Martinez-Pinedo and P. Vogel, J. Phys. **G29** (2003) 1
[20] Y. Fujita, B. Rubio and W. Gelletly, Prog. Part. Nucl. Phys. **66** (2011) 549
[21] E. Caurier, K. Langanke, G. Martínez-Pinedo and F. Nowacki, Nucl. Phys. A **653** (1999) 439
[22] E. Caurier, G. Martinez-Pinedo, F. Nowacki, A. Poves, A.P. Zuker, Rev. Mod. Phys. **77** (2005) 427
[23] K. Langanke and G. Martínez-Pinedo, Nucl. Phys. **A673** (2000) 481
[24] K. Langanke, G. Martínez-Pinedo, At. Data. Nucl. Data Tables **79** (2001) 1.
[25] A. Juodagalvis *et al* 2010, Nucl. Phys. A **848** 454
[26] A.L. Cole *et al.*, Phys. Rev. C **86** (2012) 015809
[27] S.E. Koonin, D.J. Dean and K. Langanke, Physics Reports 278 (1997) 2
[28] K. Langanke *et al.*, Phys. Rev. Lett. **90** (2003) 241102
[29] D.J. Dean, S.E. Koonin, K. Langanke, and P.B. Radha, Phys. Lett. **B399** (1997) 1
[30] D.J. dean *et al.*, Phys. Rev. **C59** (1999) 2474
[31] Q. Zhi *et al.*, Nucl. Phys. **A859** (2011) 172
[32] E. Kolbe, K. Langanke, G. Martinez-Pinedo, Phys. Rev. **C60** (1999) 052801
[33] K. Langanke, G. Martinez-Pinedo, P. von Neumann-Cosel and A. Richter, Phys. Rev. Lett. **93** (2004) 202501
[34] A. Dzhioev, A. Vdovin, V. Ponomarev and J. Wambach, Phys. At. Nucl. **74** (2011) 1162
[35] A.A. Dzhioev, A.I. Volovin, J. Wambach and V. Yu. Ponomarev, Phys. Rev. C **89** (2014) 035805
[36] H.Th. Janka and M. Rampp, Astrophys. J. **539** (2000) L33
[37] A. Mezzacappa *et al.*, Phys. Rev. Lett. **86** (2001) 1935
[38] K. Langanke *et al.*, Phys. Rev. Lett. **100** (2008) 011101
[39] A. Juodagalvis *et al.*, Nucl. Phys. **A747** (2005) 87
[40] A. Heger *et al.*, Phys. Lett. **B606** (2005)
[41] L. Huther, PhD thesis, Technische Universität Darmstadt, (2014)
[42] A. Sieverding, L. Huther, G. Martinez-Pinedo and K. Langanke, submitted to Phys. Rev. Lett. (2015)
[43] A. Byelikov *et al.*, Phys. Rev. Lett. **98** (2007) 082501
[44] T.Yoshida *et al.*, Astrophys. J. **686** (2008) 448
[45] G.J. Mathews *et al.*, Phys. Rev. **D85** (2012) 105023
[46] M.K. Cheoun *et al.*, Prog. Theor. Phys. Suppl. **196** (2012) 476
[47] T. Kajino, G.J. Mathews and T. Hayakawa, J. Phys. **G41** (2014) 044007
[48] S.G. Ryan, J.E. Norris and T.C. Beers, Astrophys. J. **523** (1999) 654
[49] K. Hirata, T. Kajita, M. Koshiba, M. Nakahata, Y. Oyama, Phys. Rev. Lett. **58** (1987) 1490.
[50] R.M. Bionta *et al.*, Phys. Rev. Lett. **58** (1987) 1494
[51] K. Langanke, P. Vogel and E. Kolbe, Phys. Rev. Lett. **76** (1996) 2629
[52] K. Scholberg, Ann. Rev. Nucl. Part. Science **62** (2012) 81
[53] T. Suzuki, S. Chiba, T. Yoshida, T. Kajino and T. Otsuka, Phys. Rev. **C74** (2006) 034307
[54] T. Suzuki, J. Phys. Conf. Ser. **80** (20119 012041
[55] T. Suzuki and M. Honma, Phys. Rev. C **87** (2013) 014607
[56] T. Suzuki and H. Sagawa, Nucl. Phys. **A718** (2003) 446c

Experimental Challenge to the νp-Process in Type II Supernovae

S. Kubono

RIKEN Nishina Center, 2-1 Hirosawa, Wako, Saitama 351-0198, Japan
Institute of Modern Physics, Chinese Academy of Sciences, Lanzhou, China
Center for Nuclear Study, University of Tokyo, 2-1 Hirosawa, Wako, Saitama 351 - 0198, Japan

kubono@riken.jp

Abstract. The νp-process is a new nucleosynthetic scenario, proposed 2006, which supposed to take place at the very early epoch of type II supernova, involving nuclear reactions of proton-rich nuclei not only with protons and alphas, but also with neutrons due to the neutrino processes. The νp-process is one of the key processes for investigating the mechanism of type II supernovae, and the process could be possibly responsible for the anomalously abundant p-nuclei around mass 90-100. Specifically, the nuclear physics problems in the νp-process were discussed in this talk including our recent experimental results with low-energy RI beams and a simulation study. Alpha cluster resonances have been identified experimentally which play a crucial role for the stellar (α,p) and (α,γ) reactions just above the alpha threshold. Neutron induced reactions in the proton-rich nuclear regions in the νp-process are also suggested to play an important role, which will discard the waiting points, and accelerate the flow to heavier nuclei. This process involves nuclear structures of very high level density at high excitation energies in neutron deficient nuclei, and both of the projectile and the target are unstable, which is a quite difficult experimental challenge in nuclear astrophysics in the coming years. Some experimental challenges are discussed.

1. The νp-process and the key stellar reactions for heavy element synthesis

Nuclear physics plays a crucial role in understanding evolution of the universe as well as various stellar phenomena. Specifically, explosive phenomena such as supernovae are considered to involve a various nuclear physics aspects which are not known well yet. They include neutrino interactions with nuclei, EOS of nuclear matter, as well as the nuclear properties in unstable nuclear regions and the reactions [1]. We discuss here the νp-process [2-4] which is considered to be one of the most challenging subjects for investigating the mechanism of type II supernovae, because it involves typical, wide-range problems for nucleosynthesis. Recently, we made a sensitivity test on the reaction rate uncertainties for the νp-process in ref. [5], identifying important nuclear reactions of (p,γ) and (α,p) as well as (n,p) and (n,α) reactions on neutron-deficient nuclei.

Journal of Physics: Conference Series **703** (2016) 012009 doi:10.1088/1742-6596/703/1/012009

The νp-process was proposed by three groups in 2006, which would take place at the very early epoch of type II supernovae in the ejecta near the inner core region [2-4]. Here, it can be proton-rich, because of the extremely high-intensity neutrino flux through the neutrino processes; $\nu_e + n \leftrightarrow p + e^-, \quad \bar{\nu}_e + p \leftrightarrow n + e^+$.

Since then, there have been some discussions about the electron fraction Ye in the very early epoch of type II supernovae mainly due to the medium effect of neutrinos, which makes Ye slightly smaller than 0.5, but soon it becomes larger than 0.5 in time [6,7]. However, the site of the νp-process may be not so much influenced by this effect because the matter density of the site for this process is not so high.

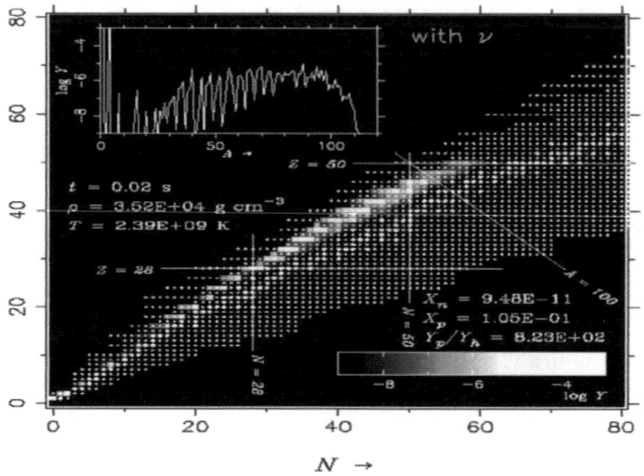

FIGURE 1. The νp-process flow predicted [4].

The νp-process has been discussed as a possible source of p-nuclei near A=90-100, which have anomalously large isotopic abundances among the p-nuclei [8]. If the νp-process runs up to this mass region, it will contribute to the p-nuclei productions. See Fig. 1. However, there are many unknown problems in nuclear physics involved in the νp-process. It includes many proton capture reactions on the neutron deficient nuclei at extremely high temperatures, but very few of them are known yet. In addition, because the site has certain fractions of neutrons and alpha particles, the νp-process involves also neutron-induced reactions as well as alpha induced reactions, which will involve high-lying excited states in proton-rich unstable nuclei. The situation can be seen in Fig. 2. The latter reactions would be relevant to α cluster resonances and the former reactions to the nuclear properties in the very high level-density region.

Apparently, there are very few reaction studies made for the (α,p) reactions, and almost none for the neutron induced reactions. Theoretical studies of the νp-process are mostly using statistical model calculations for almost all stellar reactions induced by alphas and neutrons. The statistical models in general are known that the models are not good in the light mass regions and it is also not known well in the neutron-deficient nuclear regions, because the models were studied only along the line of stability. This study should be very important.

Since the site has a significant fraction of alpha particles, alpha induced reactions would play a crucial role for the nucleosynthesis. Especially in the light mass region, (α,p) reactions will dominate in the νp-process. Here, the nuclear excitation energy region relevant for αs' is much higher than that for protons, but we expect alpha resonances there due to the cluster threshold rule, which is

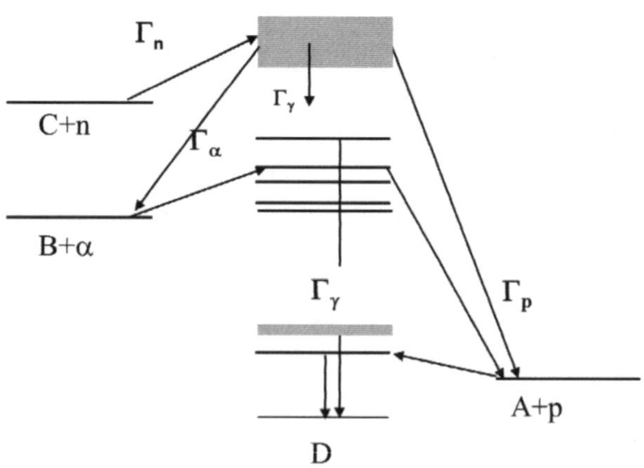

FIGURE 2. Nuclear reactions and the levels relevant to the νp-process.

naturally suggested by the Cluster Nucleosynthesis Diagram (CND) [9,10].

The nucleosynthesis in the νp-process is expected to run through neutron-deficient nuclear regions like the rp-process at high temperatures. However, an interesting consequence of neutron induced reactions in the νp-process is that the nucleosynthesis flow will skip the slow beta-decays around the waiting point nuclei along the path way of the nucleosynthesis. This would accelerate the nucleosynthesis flow to heavier mass regions, affecting the p-nucleus production more or less. Along the possible pathways, there are some critical paths to be clarified in nuclear physics in order to really establish the condition of the νp-process.

A detailed discussion on the alpha-induced stellar reactions in the νp-process will be made in sec. 2 especially for the breakout process from the pp-chain region, the discussion on the neutron induced reactions in sec. 3, and the scope of the νp-process study in summary.

2. The alpha-induced stellar reactions on the breakout process from the pp-chain region

In general, the (α,p) reactions play an important role in high-temperature hydrogen burning process in the light mass region (at A ≲ 40). The (α,p) reactions on ^{14}O [11,12] and ^{18}Ne [13] are the crucial breakout reactions from the CNO region yet to be investigated. Many other (α,p) reactions on such as ^{22}Mg and ^{30}S also will set in. The ^{21}Na(α,p) reaction is also possibly crucial for the problem of ^{22}Na nuclear gamma ray observation [14].

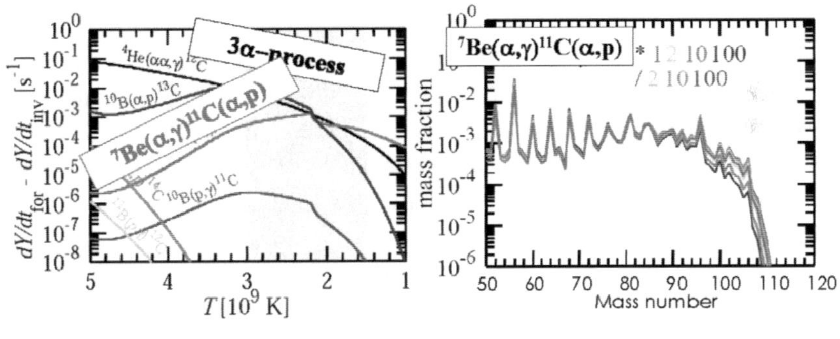

FIGURE 3. A Sensitivity test of the nuclear reactions which lead to the breakout from the pp-chain region in the νp process [5].

The (α,p) reactions in general are of course favored in the proton-rich nuclear region at high temperatures because of the positive Q values. A resonant contribution to the reaction rate of the reaction can be written as follows;

$$< \sigma v > \propto \frac{\Gamma_\alpha \Gamma_p}{\Gamma_{tot}},$$

where Γ_p and Γ_α are the proton and alpha decay widths of the resonance, respectively. This implies that one may estimate the (α,p) reaction rate by simply knowing the widths of α and p decays of the resonance. Thus, α resonant elastic scattering study is an important piece of information for the (α,p) reaction rate. Of course, α-cluster structure is also interesting for nuclear physics in proton-rich unstable nuclear region, especially near and above the α-threshold because of the cluster threshold rule [9,10], as discussed in sec. 1. Since resonance energies relevant are not large even in the νp-process environment, Γ_α cannot be so large, whereas the proton width can be often much larger, because the proton threshold is much lower in energies. Therefore, large (α,p) reaction rates can be obtained usually when Γ_α is large.

The breakout process from the pp-chain region may have possibly a common problem of nuclear physics for evolution of the first generation stars. Recent astronomical observations [15] are reporting interesting stellar objects which could be the first or the second generation stars which show a strongly enhancement of CNO elements, and very few irons together with high Sr abundance. Here, the alpha-induced reactions are expected to play a key role, which will be discussed next.

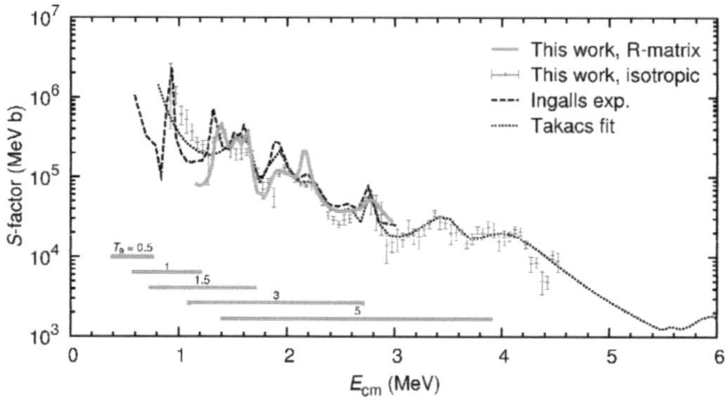

FIGURE 4. The astrophysical S factor for the $^{11}C(\alpha,p)^{14}N$ stellar reaction measured [19].

A series of experiments were made on the alpha-induced stellar reactions along the breakout process from the pp-chain region. The experiments have been made at the low-energy RI beam facility, CRIB [16,17], of Center for Nuclear Study, the University of Tokyo. There are possible three major side flows that reach CNO elements, $^7Be(p,\gamma)^8B(p,\gamma)^9C(\alpha,p)^{12}N$, $^7Be(\alpha,\gamma)^{11}C(\alpha,p)^{14}N$, and $^7Be(\alpha,p)^{10}B(\alpha,p)^{13}C$ [5]. Specifically, we discuss here our efforts on the crucial pathway, $^7Be(\alpha,\gamma)^{11}C(\alpha,p)^{14}N$.

The $^7Be(\alpha,\gamma)^{11}C$ stellar reaction was studied by measuring elastic and inelastic scattering of α + ^7Be [18]. A new resonance at 8.90 MeV, which is the third resonance above the alpha threshold, was identified for the first time, with possible spin-parity assignment of ($9/2^+$) or $3/2^+$, where the R-matrix analysis did not give a unique assignment [18]. New rotational bands were identified that have large alpha reduced widths. The reaction rates estimated through this resonance suggests about 10% increase of the total reaction rate of $^7Be(\alpha,\gamma)^{11}C$, making an assumption for the gamma width for the new resonance. Since this reaction rate is considered to have the decisive role for breakout from the pp-chain region, this gamma width needs to be investigated experimentally.

One of the crucial branching points for the breakout from the pp-chain region could be the nucleus ^{11}C, where the three processes are competing; $^{11}C(\alpha,p)^{14}N$, $^{11}C(p,\gamma)^{12}N$ and the beta decay to ^{11}B. The least known process was $^{11}C(\alpha,p)^{14}N$ reaction, which was investigated previously only by the time reverse reaction with an activation method. The $^{11}C(\alpha,p)^{14}N$ reaction study was made for the first time directly at the temperature region of interest using a low-energy, high intensity beam of ^{11}C at CRIB [19].

The cross sections were measured at Ecm= 1.0 – 4.5 MeV, which covers an effective temperature range of T = 1 - 5 $x10^9$ K. The $^{11}C(\alpha,p)^{14}N$ reaction for the low-lying excited sates in ^{14}N were also successfully observed [19], but with a large uncertainty at the lowest energy. This experiment now provides reliable reaction rate at the temperature range for the νp-process. The experiment also has confirmed the previous data of $^{11}C(\alpha,p_0)^{14}N$(g.s.) by the activation method [19], which is sensitive only to the ground state transition. The observed excitation functions of the $^{11}C(\alpha,p)^{14}N$ cross sections were characterized by individual resonances, which cannot be explained by a statistical model calculations. The experimental cross sections of the $^{11}C(\alpha,p_1)$ reaction are smaller than the statistical model prediction roughly by a factor of two. The present results demonstrate that it is important to study directly the (α,p) reaction cross sections by identifying each resonance. Some other experimental studies along the breakout process are also under way.

3. The neutron-induced stellar reaction around possible bottle neck regions

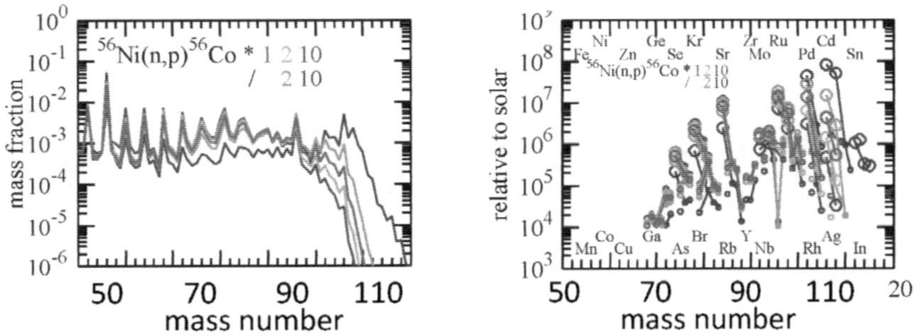

FIGURE 5. A sensitivity test of the ^{56}Ni(n,p) reaction rate for the heavy mass nucleus production in the νp-process [5].

Another interesting aspect of the νp-process is that a significant amount of neutrons also exist in the site, and thus the neutron induced reactions such as (n,p) and (n,γ) reactions would play an important role for the flow. These neutron induced reactions on proton-rich unstable nuclei are one of the most challenging subjects for experimental nuclear astrophysics. As was discussed in sec. 1, the (n,p) reactions would accelerate the process at the waiting point nuclei, which will affect eventually the production of heavy nuclei at around mass 90-100. Figure 5 shows a sensitivity test of the ^{56}Ni(n,p) rate to the heavy mass nucleus production in the νp-process [5]. The yields of A ~ 100 changes drastically. Here, the neutron-induced reactions involve high-lying states of unstable neutron-deficient nuclei, which are totally unknown yet, as discussed earlier. They are treated by the Hauser-Feshbach model. Recent theoretical works on giant resonances for wide-range unstable nuclei would improve the estimate for the rate. This subject is very interesting but it is one of the toughest challenges for nuclear astrophysics experiments. One possibility is to use an indirect method, the Trojan Horse Method (THM) with proton-rich unstable nuclear beams [21].

A successful application of the THM was made for the study of the ^{18}F(n,α)^{15}O stellar reaction using an RI beam of ^{18}F on an ^{2}H target. The stellar cross sections of ^{18}F(p,a)^{15}O studied simultaneously was already analyzed successfully [21]. Similar quality data were obtained for the ^{18}F(n,α) reaction, and the analysis is underway.

4. Summary

We discussed nuclear physics problems for the νp-process including our experimental efforts, which used low-energy RI beams from the CRIB facility. Especially, our main efforts have been placed on the (α,p) and (α,γ) stellar reactions which set in under high-temperature and high-density hydrogen burning conditions in the light mass regions. A crucial role of α resonances to the (α,p) and (α,γ) stellar reactions has been demonstrated for the first time, as suggested by the CND. Neutron induced stellar reactions on neutron deficient nuclei should be a very important but very tough experimental challenge as both the projectile and the target are unstable. An indirect method THM has been applied successfully for a test case of ^{18}F(n,α) stellar reaction study. It should be of great interest to make an innovative experimental facility like a storage ring of RI beam merging with a high-intensity neutron beam in the future for nuclear astrophysics.

Journal of Physics: Conference Series **703** (2016) 012009 doi:10.1088/1742-6596/703/1/012009

References

[1] Proc. *Origin of Matter and Evolution of Galaxies 2010*, eds. I. Tanihata, T. Shima, H.J. Ong, A. Tamii, T. Kishimoto, H. Toki, T. Kajino, and S. Kubono, *AIP Conf. Proc.* **1269** (2010).
Proc. *Origin of Matter and Evolution of Galaxies 2007*, eds. by T. Suda, T. Nozawa, A. Ohnishi, K. Kato, M. Y. Fujimoto, T. Kajino, and S. Kubono, *AIP Conf. Proc.* **1016** (2008).
S. Kubono, *Nucl. Phys.* **A693** (2001) 221.
[2] C. Frohlich, et al., *Phys. Rev. Lett.* **96** (2006) 142502.
[3] J. Pruet, R. D. Hoffman, S. E. Woosley, R. Buras, and H.-Th. Janka, *Astrophys. J.* **644** (2006) 1028.
[4] S. Wanajo, *Astrophys. J.* **647** (2006) 1323.
[5] S. Wanajo, H.-T. Janka, and S. Kubono, *Astrophys. J.* **729** (2011) 46.
[6] G. Martı´nez-Pinedo, et al. *Phys. Rev. Lett.* **109** (2012) 251104.
[7] L. F. Roberts, et al. *Phys. Rev. C* **86** (2012) 065803.
[8] M. Arnould and S. Goriely, *Phys. Rep.* **384** (2003) 1.
[9] S. Kubono, *Z. Phys. A* **349** (1994) 237.
[10] S. Kubono, Dam N. Binh, S. Hayakawa, H. Hashimoto, D. M. Kahl, Y. Wakabayashi, H. Yamaguchi, T. Teranishi, N. Iwasa, T. Komatsubara, S. Kato, Le H. Khiem, *Nucl. Phys. A* **834** (2010) 647c.
[11] M. Notani, et al., *Nucl. Phys.* **A746** (2004) 113.
[12] A. Kim, N. H. Lee, I. S. Hahn,_ J. S. Yoo and M. H. Han, S. Kubono, H. Yamaguchi, S. Hayakawa, Y. Wakabayashi, D. N. Binh, H. Hashimoto, T. Kawabata, D. Kahl, Y. Kurihara, Y. K. Kwon, T. Teranishi, S. Kato, T. Komatsubara, B. Guo, G. Bing, W. P. Liu and Y. Wang, *J. Korean Phys. Soc.*, **57** (2010) 40.
J. Hu, J. J. He, A. Parikh, S. W. Xu, H. Yamaguchi, D. Kahl, P. Ma, J. Su, H. W. Wang, T. Nakao, Y. Wakabayashi, T. Teranishi, K. I. Hahn, J. Y. Moon, H. S. Jung, T. Hashimoto, A. A. Chen, D. Irvine, C. S. Lee, and S. Kubono, *Pys. Rev. C* **90** (2014) 025803.
[13] J. J. He, L. Y. Zhang, A. Parikh, S. W. Xu, H. Yamaguchi, D. Kahl, S. Kubono, J. Hu, P. Ma, S. Z. Chen, Y. Wakabayashi, B. H. Sun, H. W. Wang, W. D. Tian, R. F. Chen, B. Guo, T. Hashimoto, Y. Togano, S. Hayakawa, T. Teranishi, N. Iwasa, T. Yamada, and T. Komatsubara, *Phys. Rev. C* **88** (2013) 012801(R).
[14] Dam N. Binh, to be published.
[15] W. Aoki, et al., *Astrophys. J.* **639** (2006) 897.
[16] S. Kubono, Y. Yanagisawa, T. Teranishi, S. Kato, Y. Kishida, S. Michimasa, Y. Ohshiro, S. Shimoura, K. Ue, S. Watanabe, and N. Yamazaki, *Eur. Phys. J.* **A13** (2002) 217.
[17] Y. Yanagisawa, S. Kubono, T. Teranishi,1, K. Ue, S. Michimasa, M. Notani, J.J. He, Y. Ohshiro, S. Shimoura, S. Watanabe, N. Yamazaki, H. Iwasaki, S. Kato, T. Kishida, T. Morikawa, and Y. Mizoi, *Nucl. Instr. Meth.* **A539** (2005) 74.
[18] H. Yamaguchi, D. Kahl, Y. Wakabayashi, S. Kubono, T. Hashimoto, S. Hayakawa, T. Kawabata, N. Iwasa, T. Teranishi, Y. K. Kwon, D. N. Binh, L. H. Khiem, and N. N. Duy, *Phys. Rev. C* **87** (2013) 034303.
[19] S. Hayakawa, to be published.
[20] P. D. Ingalls, et al., *Phys. Rev. C* **13** (1976) 524.
[21] S. Cherubini, M. Gulino, C. Spitaleri, G. G. Rapisarda, M. La Cognata, L. Lamia, R. G. Pizzone, S. Romano, S. Kubono, H. Yamaguchi, S. Hayakawa, Y. Wakabayashi, N. Iwasa, S. Kato, T. Komatsubara, T. Teranishi, A. Coc, N. de S´er´eville, F. Hammache, G. Kiss, S. Bishop, and D. N. Binh, *Phys. Rev. C* **92** (2015) 015805.

Journal of Physics: Conference Series **703** (2016) 012010 doi:10.1088/1742-6596/703/1/012010

A hitchhiker's guide to the Trojan Horse Method

R.G. Pizzone[(a)], **C. Spitaleri**[(a,b)], **S. Cherubini**[(a,b)], **G. D'Agata**[(a,b)], **G.L. Guardo**[(a,b)], **M. Gulino**[(a,c)], **M. La Cognata**[(a)], **L. Lamia**[(b)], **S.M.R. Puglia**[(a,b)], **G.G. Rapisarda**[(a,b)], **S. Romano**[(a,b)], **M.L. Sergi**[(a,b)], **R. Spartá**[(a,b)], **A. Tumino**[(a,c)]

[(a)]INFN Laboratori Nazionali del Sud, Catania, Italy
[(b)]Dipartimento di Fisica e Astronomia, Universitá di Catania, Catania, Italy
Universitá Kore di Enna, Enna, Italy

E-mail: rgpizzone@lns.infn.it

Abstract. Owing the presence of the Coulomb barrier at astrophysically relevant kinetic energies, it is very difficult, or sometimes impossible to measure astrophysical reaction rates in laboratory. This is why different indirect techniques are being used along with direct measurements. The THM is unique indirect technique allowing one measure astrophysical rearrangement reactions down to astrophysical relevant energies. The basic principle and a review of the main application of the Trojan Horse Method are presented. A step-by-step approach will be adopted in order to describe the features usually unknown to non-experts.

1. Introduction

Nuclear fusion reactions, that take place in the hot interiors of remote and long-vanished stars over billions of years, are the originof nearly all the chemical elements in the universe [1, 2, 3]. The detailed understanding of the origin of the chemical elements and their isotopes has combined astrophysics and nuclear physics, and forms what is called nuclear astrophysics. In turn, nuclear reactions are the heart of nuclear astrophysics: they influence sensitively the nucleosynthesis of the elements in the earliest stages of the universe and in all the objects formed thereafter, and control the associated energy generation (by processes called nuclear fusion or nuclear burning), neutrino luminosity, and evolution of stars. A good knowledge of the rates of these fusion reactions is essential for understanding this broad picture

In a stellar plasma the constituent nuclei are usually in thermal equilibrium at some local temperature T. Occasionally they collide with other nuclei, whereby two different nuclei can emerge from collision A+x→ c+C. The cross section σ(E) of nuclear fusion reaction A(x,c)C is of course governed by the laws of quantum mechanics where, in most cases, the Coulomb and centrifugal barriers arising from nuclear charges and angular momenta in the entrance channel of the reaction strongly inhibit the penetration of one nucleus into another. This barrier penetration leads a steep energy dependence of the cross section. It is the challenge to the experimentalist to make precise σ(E) measurements at the Gamow energy (E_G). Owing to the strong Coulomb suppression, the behavior of the cross section at E_G is usually extrapolated from the higher energies by using the definition of the smoother astrophysical factor S(E):

$$S_b(E) = E\sigma_b(E)exp(2\pi\eta) \quad (1)$$

Journal of Physics: Conference Series **703** (2016) 012010 doi:10.1088/1742-6596/703/1/012010

where $exp(2\pi\eta)$ is the inverse of the Gamow factor, which removes the dominant energy dependence of $\sigma(E)$ due to the barrier penetrability.
Although the S_b(E)-factor allows for an easier extrapolation, large uncertainties to $\sigma_b(E_G)$ may be introduced due to for instance the presence of unexpected resonances, or high energy tails of sub-threshold resonances. In order to avoid the extrapolation procedure, a number of experimental solutions were proposed in direct measurements for enhancing the signal-to-noise ratio at E_G.
Moreover, the measurements in laboratory at ultralow energies suffer from the complication due to the effects of electron screening [4, 5]. This leads to an exponential increase of the laboratory measured cross section $\sigma_s(E)$ [or equivalently of the astrophysical factor $S_s(E)$] with decreasing energy relative to the case of bare nuclei. Then, although it is possible to measure cross sections in the Gamow energy range, the bare nucleus cross section σ_b is extracted by extrapolating the direct data behavior at higher energies where negligible electron screening contribution is expected. In order to decrease uncertainties in the case of charged particle induced reactions a rather striking conclusion could be achieved: to avoid extrapolations, experimental techniques were improved. After improving measurements (at very low energies), electron screening effects were discovered. Finally to extract from direct (shielded) measurements the bare astrophysical S_b(E)-factor, extrapolation were performed at higher energy. In any case the extrapolation procedure is necessary and in consequence we find again the uncertainties problem in direct measurements.

2. The Trojan Horse Method

Alternative methods for determining bare nucleus cross sections of astrophysical interest are needed. In this context a number of indirect methods, e.g. the Coulomb dissociation (CD) [6], the Asymptotic Normalization Coefficient method (ANC)[7] and the Trojan-horse method (THM) were developed [8, 9]. For further information on the development and first principles of the method please refer to [10, 11]. The latter has already been applied several times to reactions connected with fundamental astrophysical problems such as primordial nucleosynthesis [12, 13, 14, 15, 16], lithium problem [17, 18, 19, 20, 21], light elements depletion [22, 23, 24], AGB [25, 26, 27] and Novae nucleosynthesis [28, 29, 30, 31]. THM selects the quasi-free (QF) contribution of an appropriate three-body reaction performed at energies well above the Coulomb barrier to extract a charged particle two-body cross section at energies of astrophysical interest. The idea of the THM [9, 32, 33] is to extract the cross section of an astrophysically relevant two-body reaction

$$A + x \rightarrow c + C \tag{2}$$

at low energies from a suitable chosen three-body quasifree reaction

$$A + a \rightarrow c + C + S \tag{3}$$

This is done with the help of direct theory assuming that the Trojan Horse nucleus a (the explored cases are reported in table1) has a strong $x \oplus S$ cluster structure. In many applications, this assumption is trivially fulfilled e.g. a = deuteron, x = proton, S= neutron. If the bombarding energy E_A is chosen high enough to overcome the Coulomb barrier in the entrance channel of the three-body reaction, both Coulomb barrier and electron screening effects are negligible. The polar approximation, used in the standard THM prescription has been extensively verified [34, 35] We refer to [10, 11] for further and advanced theoretical approach to the method. Instead attention will be devoted to the different steps of the data analysis required by the THM.

Journal of Physics: Conference Series **703** (2016) 012010 doi:10.1088/1742-6596/703/1/012010

Table 1. Nuclei with cluster structure which can have been used as Trojan Horse nuclei and their principal properties

	TH -nucleus	binding energy MeV	Clusters	Intercluster motion
1	d	2.225	$p-n$	l=0
2	t	6.257	$n-d$	l=0
3	^{3}He	5.494	$p-d$	l=0
4	^{6}Li	1.475	α - d	l=0
5	^{9}Be	2.467	α- ^{5}He	l=0

2.1. Identification and selection of the three-body reaction of interest

The first step in data-analysis was to identify the events related to the three-body reaction of interest for THM,$A + a \rightarrow c + C + S$ from the other ones occurring in the target. This is accomplished by studying the kinematic locus related to the above reaction and the Q-value spectrum. Coincidences between detectors aiming at c and C are examined and a typical plot of the particle energy detected in each detector is performed (see for instance fig. 4 of [34]). A narrow angular range ($\approx \pm 2^o$) is selected on both detectors and events coming from an appropriate Monte Carlo simulation, taking into account the geometrical properties of the experimental set-up as well as the features of the detectors. If a good agreement shows up, this allows further studies. Using a graphical cut which selects only the events overlapping with the Monte Carlo simulation, the Q-value spectrum is plotted (see figure 6 in [36]). A peak compatible, within the experimental errors, with the theoretical Q-value of the three-body reaction is expected. From now on only the events below the Q-value peak and and inside the kinematical graphical cut in the kinematic locus will be used for further data-analysis, in order to be relatively confident that the reaction channel $A + a \rightarrow c + C + S$ is selected.

2.2. Selection of the quasi-free mechanism

The first step after identifying the 3-body process is to investigate the reaction mechanisms involved and to separate the quasi-free (QF) contribution from any other kind of reaction mechanisms as required by the THM prescriptions. This can be done by studying, among all the available observables, the most sensitive to the reaction mechanisms which is, by far, the shape of the momentum distribution, $|\varphi(\mathbf{p_s})|^2$. According to the prescriptions in [37, 38], the momentum distribution of the third and un-detected particle is examined. This gives a major constraint for the presence of the QF mechanism and the possible application of the THM. In order to extract the experimental momentum distribution of the the undetected particle, (the spectator after the QF process is identified and selected) , $|\varphi(\mathbf{p_s})|^2_{exp}$, the energy sharing method can be applied to each pair of coincidence detectors, selecting energy intervals, ΔE_{cm}. Keeping in mind the factorization of Eq. 4, since $[(d\sigma/d\Omega)_{cm}]^{HOES}$ is nearly constant in an adequate energy interval, one can get the shape of the momentum distribution of the undetected neutron directly from the coincidence yield divided by the kinematical factor, as calculated from a suitable Monte Carlo simulation. The obtained typical momentum distribution are reported for instance in [38]. It is also compared with the theoretical distribution calculated from the Hulthen function (dashed line) with parameters taken from [39]. We can see how within the experimental errors the theoretical curve reproduces the experimental data, thus confirming the hypothesis that the neutron is acting as a spectator and that the process under investigation is a quasi-free mechanism. We only considered the s-wave since other contribution, i.e. the d-wave, were shown to be negligible [40]. According to the prescription adopted in [10] and in the standard THM approaches, only data in a limited $|p_s|$ range were chosen according to the chosen Trojan horse nucleus and used in the further analysis.

Once the experimental full width at half maximum (FWHM) Γ is obtained it has to be compared with the asymptotic theoretical value (e.g. for deuteron about 58 MeV/c) in order to highlight the presence of possible distortions. If those are present they should be taken into account as reported in [37, 38]. After this test we can stress the role of the spectator to the QF process, which constitutes a solid base for the further THM application to the studied three-body reaction for retrieving information on the binary reaction bare nucleus cross section of interest at astrophysical energies. For the following analysis only data with p_s <30 MeV/c are considered as arising from quasi-free mechanisms. This, of course, depends on the chosen Trojan Horse nucleus; the above value of p_s <30 MeV/c is typical in most cases for a l =0 relative inter cluster motion as for ^{2}H, ^{6}Li, ^{3}He.

Reaction mechanisms other than the quasi-free one should be considered as a "noise" and removed from the dataset eligible for THM application. This is done for example for the sequential mechanisms, usually occurring in the target, where an intermediate compound nucleus is formed. Examples is shown in [41]. Once the quasi-free mechanism is selected the bulk of data available for THM analysis is consistently reduced. For typical cases around 90% of data are usually rejected because they are not arising from a quasi-free process.

2.3. *Extraction of the binary reaction cross section*

In the standard THM analysis, the two body cross section is derived by dividing the experimental three-body one by the product of the kinematic factor modulated by the momentum distribution of the spectator inside the Trojan Horse nucleus [11], i.e.

$$\left(\frac{\mathrm{d}\sigma}{\mathrm{d}\Omega}\right)^{\mathrm{HOES}} \propto \frac{\mathrm{d}^3\sigma}{\mathrm{dE}_{\alpha_1}\mathrm{d}\Omega_{\alpha_1}\mathrm{d}\Omega_{\alpha_2}} / \left(\mathrm{KF}\cdot|\varphi_{\mathrm{exp}}(\mathbf{p_s})|^2\right) \tag{4}$$

Usually the factors $KF\cdot|\varphi_{\mathrm{exp}}(\mathbf{p_s})|^2$ are calculated by means of a Monte Carlo simulation, taking into account the geometrical position of the detectors. The width of the momentum distribution is set to the experimentally measured value in order to account for the distortion effects arising at low transferred momenta as discussed in [37].

The extracted $[d\sigma/d\Omega]^{\mathrm{HOES}}$ as a function of E_{cm}, corrected for the penetration factor (usually described in terms of the Regular and Irregular Coulomb functions) has to be compared to direct data in order to perform normalization at the higher energy available. This allows to calculate and compare also the astrophysical S(E)-factor, whig is extracted from the cross section according to the usual definition. This has been done in a big number of cases, many times both above and below the Coulomb barrier thus showing the strength of the method. After this necessary step the reaction-rate can be calculated according to the standard expression available in literature. Multiple example of this procedure and some results are given here for reference [10, 11, 42].

Despite several outstanding achievements like the application to radioactive ion beams or to neutron emitting or neutron induced reactions ([43, 44]), the method has still some points that should be improved, like for the necessity of normalization and the possibility of absolute measurements, the possible application to (p,γ) and (α,γ) and so on. The possibility of measuring absolute cross sections is, for example, under consideration for the next applications in order to avoid the normalization procedure to direct data.

This work was supported by the Italian Ministry of the University under Grant No. RBFR082838 and "LNS Astrofisica Nucleare (fondi premiali)".

References

[1] E.M.Burbidge, G.R.Burbidge, W.A.Fowler and F.Hoyle, Rev. Mod. Phys. **29**, (1957) 547
[2] W.A.Fowler, Rev. Mod. Phys. **56**, (1984) 149
[3] C.Rolfs and W.S.Rodney, *Cauldrons in the Cosmos* (University of Chicago Press, Chicago, 1988).
[4] H.J.Assenbaum, K.Langanke and C.Rolfs, Z. Phys.A **327**, (1987) 461
[5] F.Strieder, C.Rolfs, C.Spitaleri and P.Corvisiero, Naturwissenschaften **88**, (2001) 461
[6] G.Baur, C.A. Bertulani & H Rebel , Nucl. Phys. A **458**, (1986) 188
[7] A.M.Mukhamedzhanov and R.E.Tribble, Phys.Rev. C **59**, (1999) 3418
[8] G.Baur, Phys. Lett. B , 178, (1986) 135
[9] C.Spitaleri 1991, in *Problems of Fundamental Modern Physics, II* : Proceedings, Ed. by R. Cherubini,P. Dalpiaz, and B. Minetti (World Sci., 1991), p. 21
[10] C.Spitaleri et al., Physics of At. Nucleus **74**, (2011) 1725
[11] R.E.Tribble et al., Rep. Prog. Phys. **77**, (2014) 106901
[12] A. Rinollo et al., Nucl. Phys. A 758, (2005)146c
[13] A. Tumino et al., Phys. Lett. B **700**, (2011) 111
[14] A. Tumino et al., Astrophys.J. **785**, (2014) 96
[15] R.G. Pizzone et al., Astrophys. J. **786**, (2014) 112
[16] C. Li, et al., Phys. Rev. C **92** (2015) 025805
[17] M.Lattuada et al., Ap. J. **562**, (2001) 1076
[18] R.G. Pizzone et al., A. & A. **398**, (2003) 426
[19] R.G. Pizzone et al., A. & A. **438**, (2005) 779
[20] L. Lamia et al., A. & A. **541**, (2012) 158
[21] L. Lamia et al., Astrophys. J. **768**, (2013) 65
[22] C.Spitaleri et al., Phys. Rev. C **63**, (2001) 055801
[23] C. Spitaleri et al., Phys. Rev. C **90**, (2014) 035801
[24] L. Lamia et al., Astrophys. J. **811**, (2015) 99
[25] M. La Cognata et al., Eur. Phys. J. A**27**, (2006) 249
[26] M. La Cognata et al., Ap. J. L. **739**, (2011) L 54
[27] S. Palmerini et al., Ap. J., 764, (2013) 128
[28] M.L. Sergi et al., Phys. Rev. C 82, (2015) 032801
[29] M.L. Sergi et al., Phys. Rev. C 91, (2015) 065803
[30] S. Cherubini et al., Phys. Rev. C 92, (2015) 015805
[31] R.G. Pizzone et al., Eur. Phys. J. A in press
[32] S.Cherubini et al., Astrophys.J **457**, (1996) 855
[33] G. Calvi, Nucl. Phys. A **621**, (1997)
[34] R.G. Pizzone et al. 2011, Phys. Rev. C, 83, 045801
[35] R.G. Pizzone et al. 2013, Phys. Rev. C 87, 025805
[36] A.Tumino et al. 2008,Phys. Rev. C **78**, (2008) 064001
[37] R.G. Pizzone et al., Phys. Rev. C **71**, (2005) 058801
[38] R.G. Pizzone et al., Phys. Rev. C **80**, (2009) 025807
[39] M.Zadro et al., Phys. Rev. C **40**, (1989) 181
[40] L.Lamia,M. La Cognata, C. Spitaleri, B. Irgaziev, R.G. Pizzone, Phys. Rev. C 85, (2012) 025805
[41] L.Lamia, et al., J. Phys. G 39, (2012) 015106
[42] S. Romano, et al., Eur. Phys. Jour. A 27, (2006) 221
[43] M. Gulino et al., Phys. Rev. C 87, (2013) 012801
[44] M. La Cognata et al., Astrophys.J **805** (2015), 128

Journal of Physics: Conference Series **703** (2016) 012011 doi:10.1088/1742-6596/703/1/012011

Nuclear Astrophysics at IFIN-HH

Livius Trache

"Horia Hulubei" National Institute for Physics and Nuclear Engineering (IFIN-HH)
Bucharest-Magurele, RO-077125, Romania

livius.trache@nipne.ro

Abstract. I will present the possibilities and some results of doing nuclear astrophysics research in IFIN-HH Bucharest-Magurele. There are basically two lines of experimental activities: (1) direct measurements with beams from the local accelerators, in particular with the new 3 MV Tandetron accelerator. This facility turns out to be competitive for reactions induced by α-particles and light ions. Extra capabilities are given by the ultra-low background laboratory we have in a salt mine about 2.5 hrs. driving north of Bucharest; (2) indirect measurements done with beams at international facilities, in particular at those providing Rare Ion Beams. Completely new and unique opportunities will be provided by ELI-NP, under construction in our institute.

1. Introduction

I am glad to be again at this school! It is one of the three pillars of the European Network of Nuclear Astrophysics (ENNAS) and together with the Russbach, Austria, winter school and the Carpathian summer school in Sinaia, Romania, does an excellent service to the community. I did not want to miss coming and lecturing here. This time I did not chose the title, the organizers chose it for me. And I was glad to oblige! I will, therefore, talk about research in nuclear astrophysics (NA) at my home institute IFIN-HH, a home that I re-joined 3 years ago after a two decades absence. Besides my own group that I try to organize, there are groups with at least part of their activity motivated by NA at the accelerators the institute has. That is when we refer to experimental work. There are theoretical activities motivated by NA and I will briefly mention those, but not insist.

You (students) have learned already at this school, and I am sure you did know already, that, *grosso modo*, in experimental nuclear astrophysics we use:

1) direct measurements, that is, we measure the reactions that happen in stars at exactly the energies they happen, or as close to them as possible and
2) indirect measurements.

The first type of measurements are difficult mostly due to the very low cross sections which are typical for the very low energies available in stars. Only cases where the reactions are possible with the available projectile-target combinations (stable nuclei) can be studied so far. Particular care must be taken to get accelerators with high currents, good targets, high detection efficiencies and conditions to maintain good signal/background ratios. I will show in the next section that we can do competitively a few types of these measurements in Bucharest. The indirect methods are those methods in which we make studies using nuclear reactions at much higher energies than those available/important in the stars, typical nuclear laboratory energies 100-1000 times larger, seeking information which is later used to evaluate reaction rates at very low energies. Here there are advantages related firstly to the larger cross sections of the reactions employed and secondly to the

Journal of Physics: Conference Series **703** (2016) 012011 doi:10.1088/1742-6596/703/1/012011

fact that by using Radioactive Nuclear Beams (RNB), available lately in several laboratories, we can extend our measurements to stellar reactions which involve unstable partner(s). In IFIN-HH we do not have RNB so far, therefore this type of measurements we can only pursue at outside facilities. Important to notice is that good, reliable theoretical support is crucial in all indirect methods; I will briefly argue on that later.

2. Direct measurements

The goal of these type of measurements is to measure cross sections in the Gamow window. This means very low energies and consequently very low cross sections, a fact easily understandable for the case of reactions between charged particles (as for example (p,γ), (α,γ) or ion-ion fusion), as due to the Coulomb barrier. As such, it is no wonder that the first measurements in the Gamow window are only about one decade old! That could be done in an underground lab like that in the National Laboratory Gran Sasso and several other labs are now working with very light projectiles like p, d, or ^{3}He and different detection techniques.

In Bucharest we do not have a proton accelerator and conditions to compete in the study of proton induced reactions. But it turns out we have good conditions to study α-induced reactions and reactions between light ions important in NA. Let me elaborate on that. IFIN-HH has now 3 tandem accelerators:

- an old, but fully refurbished, 9 MV FN tandem pelletron, which is used mostly for gamma-ray spectroscopy;
- a 3 MV tandetron, installed in 2012 and which works since mid 2013 for applications like Ion Beam Analyses, ion implantation and cross section measurements [1];
- a 1 MV tandetron completely dedicated to Accelerator mass Spectrometry, used for ^{14}C dating and other such applications (like geological dating).

Since the beginning of its operation we tested the capabilities of the 3 MV tandetron for direct measurements for nuclear astrophysics. The conditions necessary for this are: appropriate energy range, stability, diversity of projectiles, high currents. We tested all of the above and found the it will be competitive. The energies are given by the terminal voltage range which is V=0.3 – 3.3 MV. It was tested to be stable over weeks of operation. It has two ion sources: a duoplasmatron and a sputter source. Na, Cs and Li charge exchange are used preferentially. These allow a diversity of beams to be provided. Table 1 shows the intensities of the beams analyzed.

Projectile	Intensity	Source
$^{1}H^{+}$	>25 μA	duoplasmatron
$^{4}He/^{3}He^{+}$	> 3 μA	
$^{11}B^{3+}$	>50 eμA	sputter
$^{12}C^{3+}$	>80 eμA	
$^{16}O^{3+}$	>80 eμA	
$^{28}Si^{3+}$	>70 eμA	
$^{31}P^{3+}$	>70 eμA	
$^{58}Ni^{3+}$	>20 eμA	
$^{63}Cu^{2+}$	>20 eμA	
$^{75}As^{2+}$	>10 eμA	
$^{197}Au^{2+}$	>80 eμA	

Table 1. Beam intensities from the 3 MV tandetron [1].

2.1. *Nuclear Astrophysics Group (NAG)*

An extra opportunity is provided by the ultra-low background laboratory the institute has in a salt mine located at about 100 km north of Bucharest-Magurele, in Slanic-Prahova [2]. We call it microBequerel. While at about 260 m below surface this is not a particularly deep mine, it has the property of being very low in natural radioactivity, due to a large distance from rocks and its compact walls. With a well shielded Ge detector a background reduction factor up to 4000 was obtained (relative to the surface background of the same unshielded detector).

We have, therefore, tested a procedure in which we irradiate targets in Magurele, then transfer them in Slanic and measure them [3]. Obviously this procedure will not work for cases where the resulting activity after irradiation has half-lives much shorter than the transfer time of about 2.5 hrs.

One case ideal for the test of the procedure we proposed together with colleagues from China (IMP Lanzhou and CIAE Beijing): the reaction $^{13}C+^{12}C$. The motivation of the experiment, the setup of 2014 and its preliminary results were presented at this school by one of our students and is included in this volume [4], please see it for details. You can see there pictures of the accelerator and of the salt mine. I will only mention here that only one reaction channel leads to radioactivity (one-proton evaporation), ^{24}Na which has $T_{1/2}$=15.0 h, excellent for the procedure we used: one day of irradiation, transfer to Slanic in 2.5 hrs and about one day de-activation measurement there, during the irradiation of the next target, and so on… With these we could reach (measurements in Sept.-Oct. 2014 and Oct. 2015) cross sections of the order of tens of picobarns, about 100 times more sensitive than any measurement done before. In the latter experiment we also measured prompt gamma-rays (as far down in projectile energy as we could) to assess the contribution of the other open reaction channels. Another condition, which did I not mention so far, is the availability of good detection devices. We have several high resolution, high efficiency HpGe detectors (100-120 % relative efficiency), many large Si detectors and we are building a large (80 cm diameter) reaction chamber. I mention that the tandems accelerators are internationally open facilities, with PAC meeting annually and are transnational access facilities under the ENSAR2 project.

2.2. Other activities in nuclear astrophysics

There are also groups in the Department of Nuclear Physics (DFN) of IFIN-HH using beams from the 9 MV tandem for nuclear astrophysics. They measured (α,γ) reactions at relatively low bombarding energies on medium mass targets, having in mind to obtain data for modeling the nucleosynthesis of the *p-nuclei*. One such example is presented at this school by Andreea Oprea [5]. Detailed measurements, down to energies close to the Gamow window, can provide data to determine optical potentials for α-particles at low energies and radiation strength functions. The experiments typically measure prompt or activation gamma-rays as the department is reach in gamma-ray detectors and in experience in using them. Experimental measurements are complemented by theory efforts to provide systematics of α-particle optical potentials at low and very low energies by Vlad and Marilena Avrigeanu of the same department (see for example [6] and references therein).

3. Indirect measurements at outside facilities

This type of measurements involve radioactive nuclear beams, as I said before, and they are planned or actually done at outside facilities. I will not talk generally about indirect methods in NA here, I was doing it at earlier editions of the school, they were discussed by prof. Carlos Bertulani a few days ago. I will only mention two of them, pursued by my NAG group.

3.1. Coulomb and nuclear breakup of ^{9}C at RIBF of RIKEN

We have proposed earlier to use breakup reactions to extract Asymptotic Normalization Coefficients and from there radiative proton-capture rates [7]. We were talking about nuclear breakup. Coulomb breakup was proposed long before and used as a reliable indirect method in NA [8, 9]. The current knowledge of the rate of the $^{8}B(p,\gamma)^{9}C$ reaction in stellar conditions is contradictory at best and there is no hope to determine it, now or ever, by other means than by indirect methods. This reaction gives a possible path to the hot *pp* chain *pp-IV* at high temperatures and away from it toward a rapid alpha process *rap I* at high temperatures and densities and therefore is important in understanding nucleosynthesis in super-massive hot stars in the early universe, including possible bypasses of the 3α-process. Our best hope at the determination of the astrophysical factor S_{18} at low energies is by using the $^{9}C\rightarrow^{8}B+p$ breakup. We proposed to use a combination of nuclear and Coulomb dissociation measurements using the SAMURAI spectrograph of RIBF at RIKEN (on a light target – Be or C and on a heavy target – Pb, respectively) at two energies (100 and 300 MeV/nucleon) to extract structure information which will allow to evaluate the radiative proton capture cross section at low energies and from there the reaction rate. We proposed an exclusive study of the reaction, which may allow a better understanding of the reaction mechanism (proposal NP1412-SAMURAI29R1, approved by PAC Dec. 2014). The high probability of two-proton breakup $^{9}C\rightarrow^{7}Be+2p$ for this projectile makes it a good case to learn about the complex reaction mechanisms involved. Such an understanding may allow for a better theoretical description of the reactions and more accurate calculations of the momentum and angular distributions, a crucial step in using indirect methods for nuclear astrophysics.

Not negligibly, the reaction proposed is the easiest among the p-HI experiments being planned at this point with SAMURAI and will be a good start for the use of the Si detector system in front of SAMURAI as it results in a smaller dynamical range and an easier particle identification and less kinematic focus. Part of the preparations for these experiment, actually for a whole set of proton breakup experiments proposed at SAMURAI was to build the detection system in front of the spectrograph which involves many channels (upward of 1024) and with the capability to measure energies on a very broad dynamic range (from 200 keV to 600 MeV). An approach based on ASICs was proposed, developed and tested at the HIMAC medical facility in Chiba, Japan. I do not detail that multi-year work here.

3.2. β-delayed proton decay with AstroBox2 at Texas A&M University

Last edition of the school I discussed the design of a new type of detector to measure β-delayed proton emission, using a detection scheme that involved the stopping of the p-emitting nuclei in the middle of a gas detector after their production and separation with the MARS recoil separator of the Cyclotron Institute at Texas A&M University. The β-delayed protons emitted produce ionization in gas and the resulting charges are directed and then amplified with very efficient and good resolution devices called micromegas. The detector was called Astrobox1 [10] and we have shown its use for measuring very low energy protons, as well as the connection between these measurements and the determination of astrophysical reaction rates for proton induced reaction rates dominated by resonances. Its main advantage over very the thin Si detectors (45-65 μm) we have used in the first stage of this project is that it is less sensitive to betas emitted always in the first stage of the decay, which gave a large continuous background in Si at low energies, exactly in the energy range of interest for us (protons of 150-500 keV). The sensitivity is down to around 10^{-4} (p-branching ratios). Since, we have worked, with our colleagues at CEA/IRFU Saclay, TAMU and CERN on an improved version of this detection scheme. The detector, dubbed Astrobox2, was built in this transcontinental collaboration and the first test measurements were done at the Cyclotron Institute, Texas A&M University, in College Station, TX in April 2015. The active part of the detector, the micromegas was built for us by a group at CERN, the inventors of such devices. The body of the detector was designed in Texas by dr. A. Saastamoinen. Its design and realization was actively pursued by us and by our collaborator from CEA/IRFU Saclay, France, dr. E. Pollacco. Schematically is shown in Figure 1. The main difference from AstroBox1 is that it does not have anymore a cylindrical symmetry (on an axis perpendicular to the beam), but is more appropriate to the geometry of the beam and its stopping in the gas of the detector. Another difference is that it has 29 separated pads and correspondingly 29 signals, compared with 3 (or 5) only for AstroBox1 (AB1).

The tests that were done, were:

- Off-beam tests using ^{55}Fe and ^{241}Am sources
- In-beam commissioning of the detector using a ^{25}Si radioactive beam separated by the MARS spectrometer. The radioactive specie we used, ^{25}Si, is very appropriate for a test of a new proton detector, as it is a good, well known, β-delayed proton emitter with a large βp-branching. It was produced at a reasonable rate from a primary beam of ^{28}Si at 40 MeV/nucleon on a ^{27}Al solid target. The result was actually a cocktail of secondary beams, a benefit for the identification of the beam in AB2.
- In-beam measurements for the ^{23}Al secondary beam, which is the main focus of the physics for these measurements.

The same scheme of measurements was used in 2012 on a test of AB1. The first two parts of the tests went very well, and the commissioning of the AB2 detector was a success. So appeared to be the last measurement, for which we reserved the last 4 days of the experiment. The primary beam of ^{24}Mg at 45 MeV/nucleon was delivered successfully to us on Friday April 24, and the next day the secondary beam of ^{23}Al was selected and was stopped in the middle of the AB2. The data need still be fully analyzed to see if the reconstruction of all signals from adding the adjacent pads improved the statistics to the point we need.

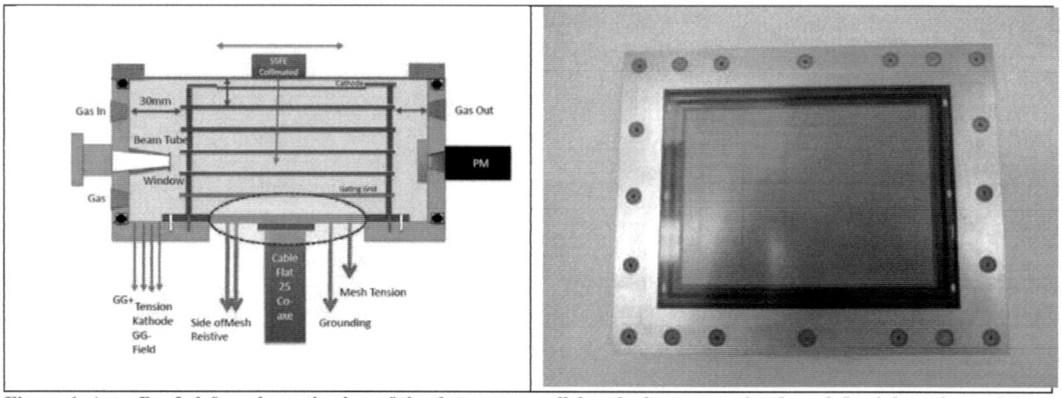

Figure 1. AstroBox2: left - schematic view of the detector, parallel to the beam entering from left; right - photo of the micromegas plate.

3.3. Theory for indirect methods in NA

We have a long-term program to understand and describe nucleus-nucleus collisions in terms of one interaction potential, the optical model potential (OMP). A good understanding of all phenomena occurring in the elastic nucleus-nucleus scattering, which are used typically to extract OMP, and the interpretation of the origin of different aspects, including the well know potential ambiguities, are of crucial importance for finding and justifying the procedures used for predicting nucleus-nucleus OMP in the era of radioactive nuclear beams (RNB), including ours based on double folding [11]. The reliability of these potentials is crucial in the correct description of a number of reactions involving RNBs, from elastic to transfer, to breakup, at energies ranging from a few to a few hundred MeV/nucleon. Of particular interest for us is to support the absolute values of the calculated cross sections for reactions used in indirect methods for nuclear astrophysics [13]. In this framework, we paid and pay particular attention to obtain OMP using double folding and to finding systematics for the re-normalization coefficients that appear as needed in quasi-all cases. It is extremely important for the use of indirect methods to be able to obtain and rightfully claim good absolute values for the calculated cross sections for various phenomena (including ^{9}C breakup, e.g.), see Refs. 12-14 for the most recent results.

4. ELI-NP

The most important and the most notable news from IFIN-HH in the last 3 years is ELI-NP. This is a large European project being built by our institute, in our institute and is designed to be an independent European institution in a few years. It consists of two 10 PW lasers and a gamma-beam system capable of delivering brilliant and mono-energetic gamma beams up to 19-20 MeV. These powerful lasers will make possible a nuclear physics program that has many potentially unique capabilities. Is being described in the talk by dr. O. Tesileanu here, including its program in nuclear astrophysics, which is developed in collaboration with scientists from Europe and the whole world, including many from LNS Catania. I will list here only one idea that may not have been included in that talk. It is expected that at ELI-NP stellar plasma conditions will be produced for long enough periods of time to have nuclear reactions in equilibrium at large temperatures T. There will be many problems, like how to realize those stellar plasmas, how to characterize them, how to obtain and measure the signals, but among other things, they may provide the unique opportunity to measure reaction rates for capture reactions on excited states and without electron screening [15]. Impossible to realize otherwise in any current laboratory!

And, by the way: one of the best things for the future users of ELI-NP is that there is already a 1 PW laser running on our campus, not in IFIN-HH, but in the laser institute nearby (INFLPR).

Acknowledgements
This work was supported by Romanian National Authority for Scientific Research and Innovation under grants PNII-ID-PCE-2012-4-0299 and PN-09-37-01-07.

References

[1] I. Burducea, M. Straticiuc, D.G. Ghita, D.V. Mosu, C.I. Calinescu, N.C. Podaru, D.J.W. Mous, I. Ursu, N.V. Zamfir, **Nuclear Instruments and Methods in Physics Research B 359**, 1219, 2015.
[2] R. Margineanu, C. Simion, S. Bercea, O.G. Duliu, D. Gheorghiu, A. Stochioiu, M. Matei, **Applied Radiation and Isotopes 66**, 1501-1506, 2008.
[3] D. Chesneanu et al. in "Exotic Nuclei and Nuclear/Particle Astrophysics (V). From Nuclei to Stars", *Proceedings CSSP14*, L. Trache, D. Chesneanu and C.A. Ur (eds.), **AIP Conf. Proc. Series, 1645**, 311 (2015).
[4] D. Tudor et al., these Proceedings
[5] A. Oprea et al., these Proceedings
[6] V. Avrigeanu and M. Avrigeanu, **Phys. Rev. C 91**, 064611 (2015)
[7] L. Trache, F. Carstoiu, C.A. Gagliardi and R.E. Tribble, **Phys. Rev. Lett. 87**, 271102 (2001).
[8] G. Baur, C.A. Bertulani and H. Rebel, **Nucl. Phys. A458**, 188 (1986).
[9] T. Motobayashi et al., see for example **Nucl. Phys. A682**, 345c (2001).
[10] E. Pollacco, L. Trache et al., **Nucl. Instr. Meth. Phys. Res. A 723**, 102 (2013).
[11] L. Trache, A. Azhari, H. L. Clark, C. A. Gagliardi, Y.-W. Lui, A. M. Mukhamedzhanov, R. E. Tribble, F. Carstoiu **Phys. Rev. C 61**, 024612(2000).
[12] T. Al-Abdullah, F. Carstoiu, X. Chen, H. L. Clark, C. A. Gagliardi, Y.-W. Lui, A. Mukhamedzhanov, G. Tabacaru, Y. Tokimoto, L. Trache, R. E. Tribble, and Y. Zhai, **Phys. Rev. C 89**, 025809 (2014).
[13] M. McCleskey, A. M. Mukhamedzhanov, L. Trache, R. E. Tribble, A. Banu, V. Eremenko, V. Z. Goldberg, Y.-W. Lui, E. McCleskey, B. T. Roeder, A. Spiridon, F. Carstoiu, V. Burjan, Z. Hons, and I. J. Thompson, **Phys. Rev. C 89**, 044605 (2014).
[14] T. Al-Abdullah, F. Carstoiu, C. A. Gagliardi, G. Tabacaru, L. Trache, and R. E. Tribble, **Phys. Rev. C 89**, 064602 (2014).
[15] L. Trache in http:// http://eurisol.ifj.edu.pl/eurisol_krakow_proceedings.pdf , pp. 101-104.

RIB Production at LNL: the EXOTIC Facility

Marco Mazzocco

Dipartimento di Fisica e Astronomia, Università di Padova, via F. Marzolo 8, I-35131 Padova, Italy

INFN-Sezione di Padova, via F. Marzolo 8, I-35131 Padova, Italy

E-mail: marco.mazzocco@pd.infn.it

Abstract. Nuclear reactions involving radioactive isotopes are extremely relevant in several astrophysical scenarios, from the Big-Bang Nucleosynthesis to Supernovae explosions. In this contribution the production of Radioactive Ion Beams (RIBs) by means of the in-flight technique is reviewed. In particular, the use of direct reactions in inverse kinematics for the production of light weakly-bound RIBs by means of the facility EXOTIC at INFN-LNL (Italy) will be described in detail.

1. Introduction

Radioactive nuclei are involved in several processes of astrophysical interest, ranging from nuclear reactions occurred in the early universe to reaction mechanisms still taking place in stars or in cosmic explosions [1]. Beams of radioactive isotopes provide us the possibility of investigating these processes on Earth. Unfortunately, due to the present-day limited intensity of Radioactive Ion Beams (RIBs), these studies still suffer, in most cases, of low statistical accuracy.

Considerable effort has been currently put by the worldwide Nuclear Physics community for the construction of second generation RIB factories. There exist two main techniques for the production of RIBs: the In-Flight (IF) method and the Isotope Separation On Line (ISOL). In a very schematic description, the IF approach uses heavy ion beams impinging on thin targets. The reaction products retain most of the projectile initial velocity and are in-flight selected by proper combinations of electromagnetic fields before being delivered to the experimental areas. In the other side, the ISOL technique employes either light or heavy ion beams impinging on thick targets kept at high temperature. The reaction products are thus created in a sort of ion source. Then they thermally diffuse out of it, they are mass selected and finally post-accelerated before being delivered to the experimental halls. Both methods have advantages and drawbacks. For instance, IF beams usually require rather faster separation times (in the order of a few μs), but have poor optical properties, i.e. large emittance. On the contrary, ISOL beams need longer selection times (in the range between a few ms to a second), but the outcoming beams have very good optical properties.

In this contribution, we will initially review the characteristics of the IF production methods (Sec. 2), then we will describe in detail the RIB production by means of direct reactions in inverse kinematics (Sec. 3). As an application, we will present the RIBs produced so far by the facility EXOTIC [2] at INFN-LNL (Italy) (Sec. 4). Some concluding perspectives will finally be given in Sect. 5.

2. In-Flight production method

The main positive aspect of the IF methods is the use of (relatively) thin production targets (up to a few mg/cm^2). In this way, the reaction products retain a large fraction of the initial projectile velocity and, due to linear momentum conservation, they have a kinematic (forward) focusing at small angles around the beam direction. A fragment separator can thus be located exactly along the primary beam axis. The selection of the RIB of interest is performed by means of a proper combination of electromagnetic fields and atomic interaction and is generally a process quick (the transit time through a spectrometer is at maximum of the order of a few μs) and independent from the chemical properties of the radioisotopes under production. Moreover, IF RIBs usually do not require the design of post accelerators and consequently IF facilities result to be somewhat less costly than ISOL-based laboratories. However, IF RIBs typically have poor optical quality, in terms of longitudinal and transverse emittance and beam spot on target.

There exist three main IF production techniques: (*i*) projectile fragmentation, (*ii*) projectile fission and (*iii*) direct processes in inverse kinematics. To be effective the first two processes require a projectile energy of a few tens MeV/u and thus a cyclotron, if not a synchrotron, is needed as driver accelerator. The last method is less demanding in terms of beam energy, since it employs direct processes, whose cross sections reach the range of hundreds or thousands mb already at a bombarding energy of a few MeV/u. In this case, even old machines like electrostatic tandem accelerators can be used as primary driver.

3. Direct reactions in inverse kinematics

This IF method employs inverse kinematics reactions (i.e. heavy ion beams of light targets), such as inverse (p,n), (d,n), (d,p) or (^{3}He,n) processes, preferably with negative Q_{values}. The large linear momentum carried by the heavy projectile implies that the reaction products are emitted in rather narrow cones around the primary beam direction.

In a standard two-body nuclear reaction in inverse kinematics:

$$p + t \rightarrow p_{like} + t_{like} \qquad (1)$$

where p, t, p_{like} and t_{like} are the projectile, target, projectile-like and target-like fragment, respectively, the maximum opening (θ_{max}) of the emission cone for projectile-like particles is given by the formula:

$$\theta_{max} = \arccos\sqrt{-\frac{(M_{t_{like}} + M_{p_{like}})[M_{t_{like}}Q_{value} + (M_{t_{like}} - M_p)E_{beam}]}{M_p M_{p_{like}} E_{beam}}} \qquad (2)$$

being M_p, M_t, $M_{p_{like}}$ and $M_{t_{like}}$ the projectile, target, projectile-like and target-like masses, respectively, E_{beam} the beam energy (in the laboratory reference frame) and $Q_{value} = M_p + M_t - (M_{p_{like}} + M_{t_{like}})$.

The preferred target materials would be hydrogen or helium isotopes. Unfortunately, the are no He-compounds, while H-rich compounds cannot withstand the thermal stress induced by a high intensity primary beam. To circumvent these problems, the use of gas targets, often cooled down to liquid nitrogen temperatures, with metallic windows is generally undertaken.

In the following subsections, we will describe in detail the dependence on beam energy, reaction Q_{value} and projectile mass M_p of the maximum emission angles for projectile-like fragments produced in (p,n), (d,p), (d,n) and (^{3}He,n) reactions.

3.1. (p,n) reactions

Figs. 1 and 2 show that the maximum opening of the emission cones for projectile-like fragments produced in (p,n) charge-exchange reactions decreases as the projectile mass increases. This

Journal of Physics: Conference Series **703** (2016) 012012 doi:10.1088/1742-6596/703/1/012012

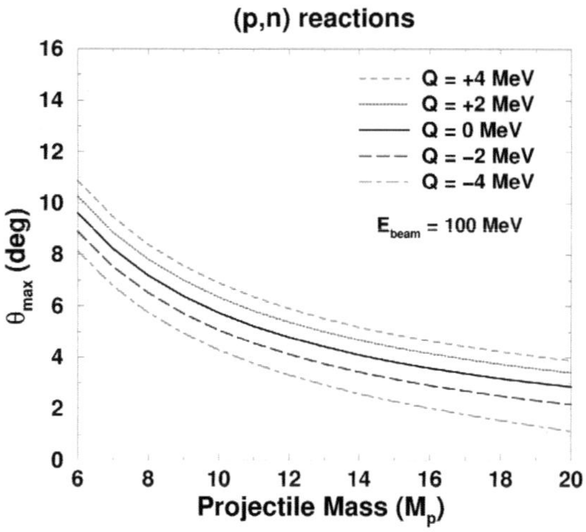

Figure 1. Maximum emission angles (θ_{max}) for projectile-like fragments produced in (p,n) reactions as a function of the projectile mass (M_p). Different lines correspond to reaction Q_{values} (Q) in the energy range from -2 to +2 MeV. The beam energy (E_{beam}) was arbitrarily set to 100 MeV.

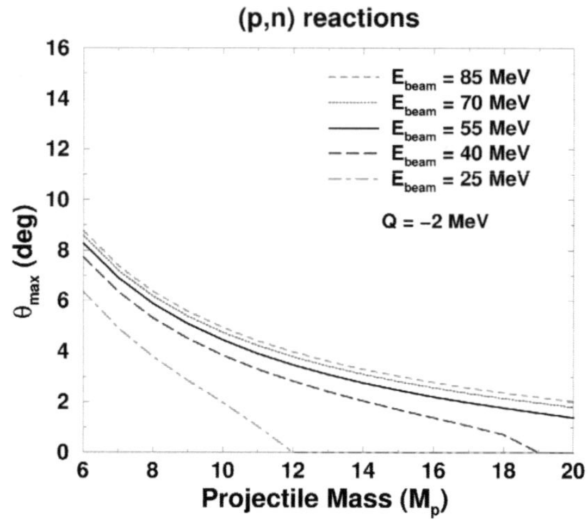

Figure 2. Same as Fig. 1. Different lines correspond to beam energies (E_{beam}) ranging from 25 to 85 MeV. The reaction Q_{values} (Q) was arbitrarily set to -2 MeV.

feature is clearly related to larger linear momentum carried by heavier projectiles. We also notice that the emission cones get significantly narrower for negative Q_{values} (Fig. 1) and for smaller primary beam energies (Fig. 2). In particular, we can observe in Fig. 2 that when the beam energy is equal to 25 and 40 MeV, the opening of the emission cones vanishes for projectile mass larger than 12 and 19, respectively. The plotted curves were indeed calculated for an hypothetical reaction with a Q_{value} of -2 MeV and, in the laboratory frame, there is a threshold energy, given by the formula:

$$E_{beam,min} = -Q_{value}\frac{M_p + M_t}{M_t} \tag{3}$$

for a negative Q_{value} reaction to occur.

3.2. (d,p) and (d,n) reactions

Figs. 3 and 4 display the maximum emission angles for projectile-like fragments produced either in (d,p) or (d,n) reactions. In these cases, either a loosely-bound neutron or proton is transferred from the deuterium target to the projectile, generating a n-rich or a p-rich RIB, respectively. The increased target mass implies that the openings of the emission cones are somewhat larger with respect to RIBs produced via (p,n) reactions.

3.3. (^{3}He,n) reactions

We finally considered (^{3}He,n) reactions, where two protons are transferred from the gas target to the projectile. These reactions allow producing p-rich RIBs two mass units apart from the valley of β-stability. In this case the maximum emission angle can reach values as large as θ_{max} = 13°, as for the production of a ^{8}B secondary by means of the reaction ^{6}Li + ^{3}He → ^{8}B + n (Q_{value} = -1.97 MeV).

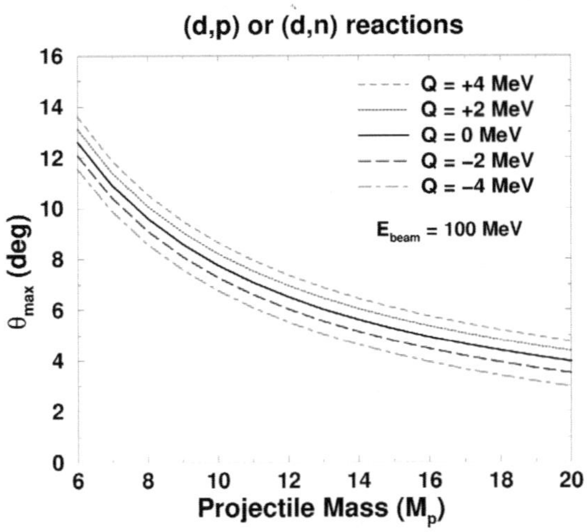

Figure 3. Same as Fig. 1, but for projectile-like fragments produced in (d,p) or (d,n) reactions.

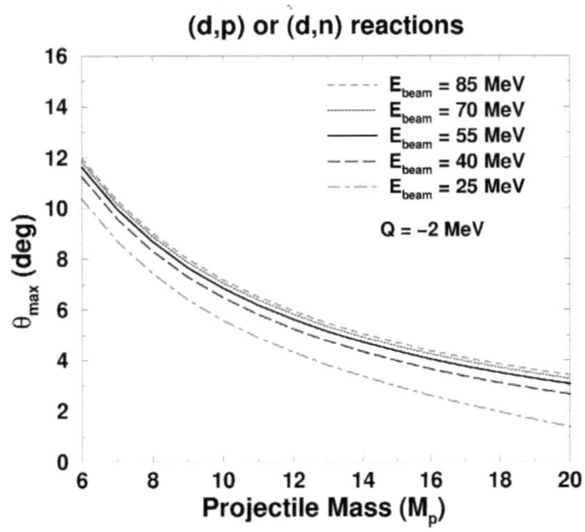

Figure 4. Same as Fig. 2, but for projectile-like fragments produced in (d,p) or (d,n) reactions.

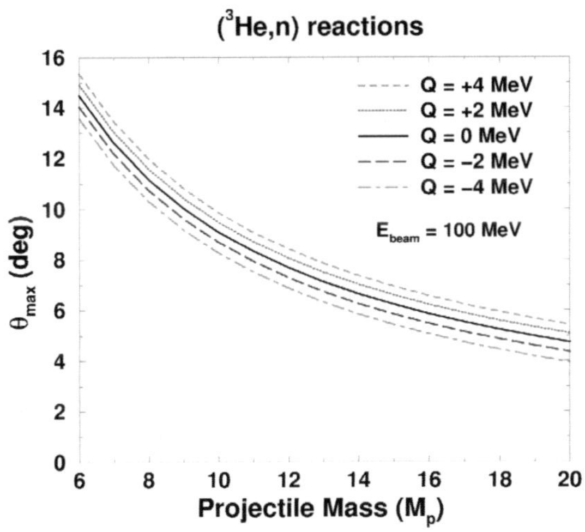

Figure 5. Same as Fig. 1, but for projectile-like fragments produced in (^{3}He,n) reactions.

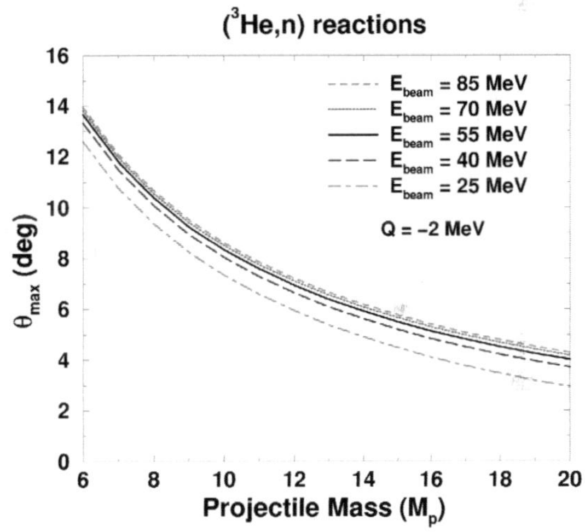

Figure 6. Same as Fig. 2, but for projectile-like fragments produced in (^{3}He,n) reactions.

3.4. Reaction Q_{value}

Fig. 7 summarizes the reaction Q_{values} for RIBs which can be produced in two-body processes in inverse kinematics induced on a hydrogen, deuterium and ^{3}He gas targets. The data are plotted as a function of the primary beam mass. ^{9}Be, $^{20-22}$Ne and ^{36,38,40}Ar beams, which cannot be delivered by the LNL-XTU tandem accelerator, were excluded from the graph. We can observe that all (p,n) charge exchange reactions have negative Q_{values}. Generally speaking, this a quite favorable circumstance, since the RIB particles are emitted much closer to the beam axis and ion-optical elements (e.g. quadrupoles and dipoles) with smaller openings could be employed without compromising too much the transmission of the fragment separator. However, we should keep in mind, that the trick of using inverse kinematics reactions for the RIB IF production is

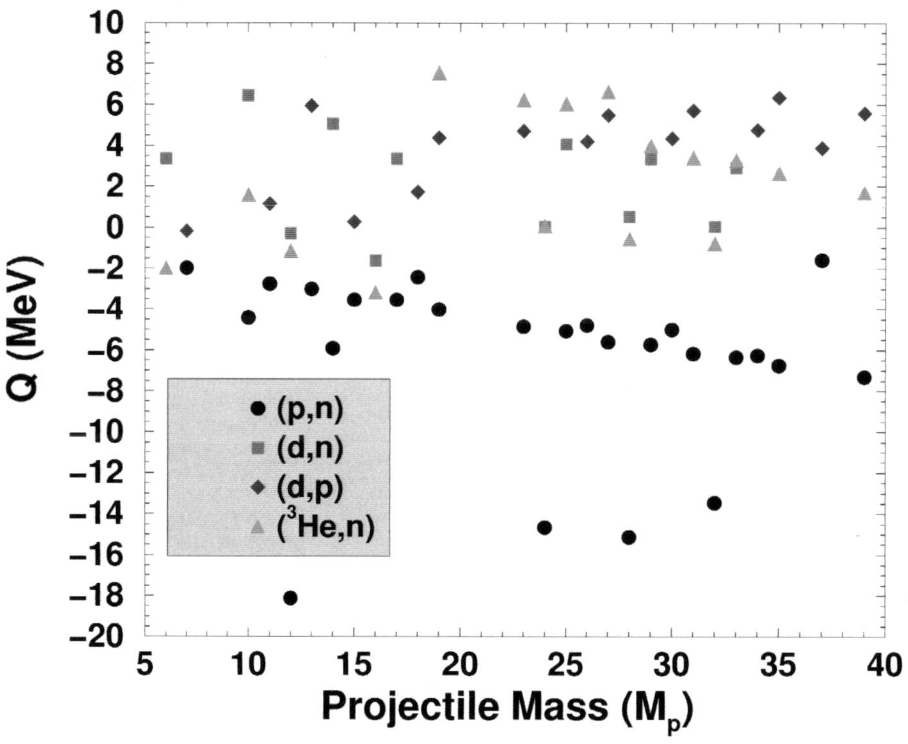

Figure 7. Reaction Q_{values} for RIBs which can be produced by means of (p,n), (d,n), (d,p) and (^{3}He,n) reactions. The data were plotted as a function of the projectile mass. Beams, such as ^{9}Be, $^{20-22}$Ne and ^{36,38,40}Ar, which cannot be delivered by the LNL-XTU tandem accelerator were not included in the figure.

not really effective when the reaction Q_{value} is much smaller than -5 MeV and for RIB masses heavier than 25 amu, as the minimum energy required for the production reaction to occur in the laboratory frame may easily exceed the capabilities of an electrostatic accelerator, such as the LNL 15 MV tandem.

In addition, we notice that (d,p) reactions usually have positive Q_{values} (the only notable exemption is the reaction ^{7}Li + d employed for the production of a ^{8}Li RIB), whereas several (d,n) reactions might be good candidates for the production of p-rich RIBs. Finally, a few (^{3}He,n) reactions have negative Q_{values}, especially for light projectiles masses, and can be possibly employed for the production, for instance, of ^{8}B, ^{14}O and ^{18}Ne RIBs.

4. RIB production at EXOTIC

The facility EXOTIC [2] was commissioned at INFN-LNL in the years 2003-2004 and was upgraded in 2012 [3]. The production target consists of a 5-cm long gas cell doubly-walled with 2.2-μm thick havar windows. The target station can operated either at room temperature or can be cooled down to liquid nitrogen temperature ($\approx$ 90 k). The standard operating gas pressure is 1 bar and the windows were tested up to 1.4 bar. The ion optical lay-out is made up by eight elements: a first quadrupole triplet (Q1-Q3), a 30°-dipole magnet (DM), a 1-m long Wien (velocity) filter (WF) and a second quadrupole triplet (Q4-Q6). The entire facility is about 8.4 m long.

The secondary beam selection is performed in two stages. At first, the DM selects particles

Journal of Physics: Conference Series **703** (2016) 012012 doi:10.1088/1742-6596/703/1/012012

with the same (within the acceptance of the magnet) magnetic rigidity, $B\rho = \frac{mv}{q}$, being B the magnetic field of the DM, ρ the radius of curvature and m, v and q the particle mass, velocity and charge, respectively. According to the basic laws of electromagnetism, different ion species characterized by the same magnetic rigidity have the following kinetic energies, E, and velocities, v:

$$E = \frac{q^2}{2m}(B\rho)^2 \tag{4}$$

$$v = \frac{q}{m}(B\rho) \tag{5}$$

We can immediately observe that the kinetic energy depends quadratically on the magnetic rigidity, whereas the velocity scales linearly with $B\rho$. As an example, Figs. 8 and 9 illustrate the behavior of the energy and velocity, respectively, as a function of the magnetic rigidity in ^{17}F RIB production by means of a (p,n) reaction. We considered in the figure non only the $^{17}F^{9+}$ secondary beam, but also four different charge states of the scattered beam, i.e. $^{17}O^{8+,7+,6+,5+}$, and three charge states of a possible contaminant beam, i.e. $^{16}O^{8+,7+,6+}$.

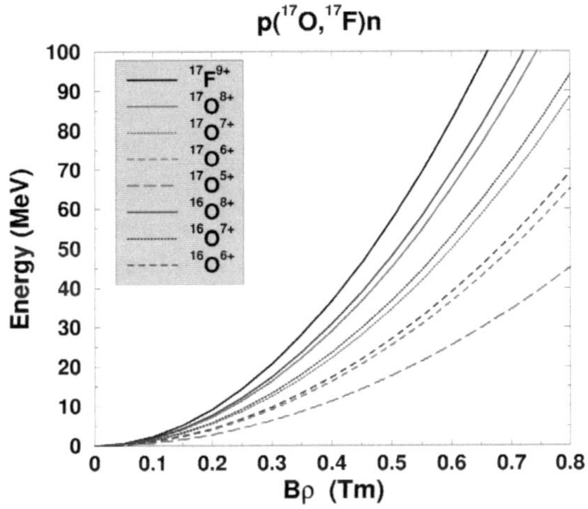

Figure 8. ^{17}F RIB production: kinetic energy as a function of the magnetic rigidity for the following beams: $^{17}F^{9+}$ (black curve), $^{17}O^{8+,7+,6+,5+}$ (red curves) and $^{16}O^{8+,7+,6+}$ (blue curves).

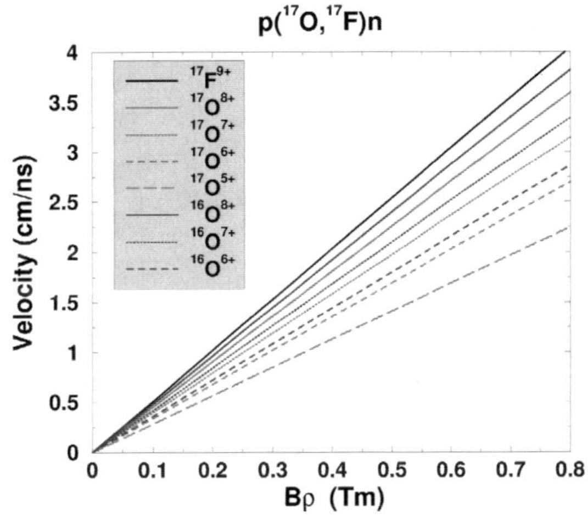

Figure 9. ^{17}F RIB production: velocity as a function of the magnetic rigidity for the following beams: $^{17}F^{9+}$ (black line), $^{17}O^{8+,7+,6+,5+}$ (red lines) and $^{16}O^{8+,7+,6+}$ (blue lines).

Fig. 8 shows that the parabolas describing the energies of the different ion species can essentially be grouped according to the particle charge state. The most energetic beam is the only one with charge state 9+ (i.e. $^{17}F^{9+}$), then we have beams with charge state 8+ ($^{17}O^{8+}$, $^{16}O^{8+}$ and, not shown, $^{17}F^{8+}$) with smaller energies, then 7+, 6+ and so on. We can also identify a fine structure within these groups, since lighter masses, according to Eq. 4, have higher kinetic energies.

The selection provided only by the DM would produce a cocktail beam with the desired RIB together with several different charge states of the scattered beam and, possibly, some other contaminant beams. A typical example of such a circumstance is given in the upper panel of Fig. 2 in Ref. [4], where one clearly sees a small bump related to the $^{17}F^{9+}$ RIB, a huge peak due to the $^{17}O^{8+}$ scattered beam and then the other ^{17}O charge states with decreasing intensity.

For this reason, the facility was additionally equipped with a 1-m long WF. In this region we have the superposition of a magnetic field and a perpendicular electrostatic field, E, in such a way that the Lorentz force ($\vec{F_L} = q\vec{v}\times\vec{B}$) would be opposite to the electrostatic force ($\vec{F_E} = q\vec{E}$). During an experiment, a proper tuning of both electric and magnetic field of the WF ensures an overall suppression of all contaminant beams and helps achieving a nearly 100-% purity for RIBs produced by means of (p,n) and (d,p) reactions. Some examples are provided by the lower panel of Fig. 2 in Ref. [4] and by Fig. 1 in Ref. [5] for the ^{17}F and ^{7}Be secondary beam production, respectively. Similar scenarios were also observed for the production of ^{15}O and ^{10,11}C radioactive beams. In all these cases beam purities as good as 99 % were achieved.

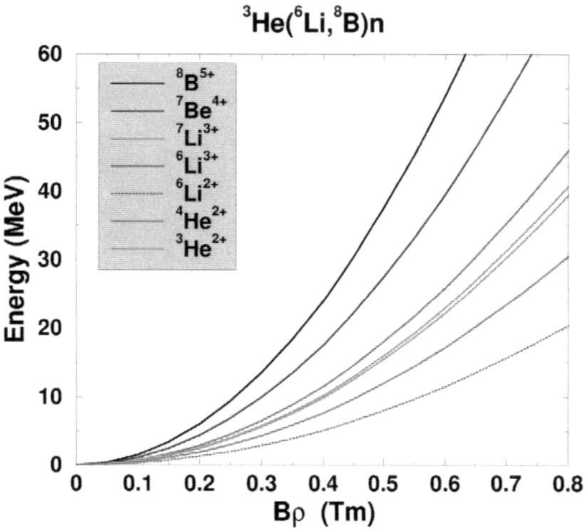

Figure 10. ^{8}B RIB production: kinetic energy as a function of the magnetic rigidity for the following beams: ^{8}B^{5+} (black curve), ^{7}Be^{4+} (red curve), ^{7}Li^{3+} (green curve), ^{6}Li$^{3+,2+}$ (red curves), ^{4}He^{2+} (pink curve) and ^{3}He^{2+} (yellow curve).

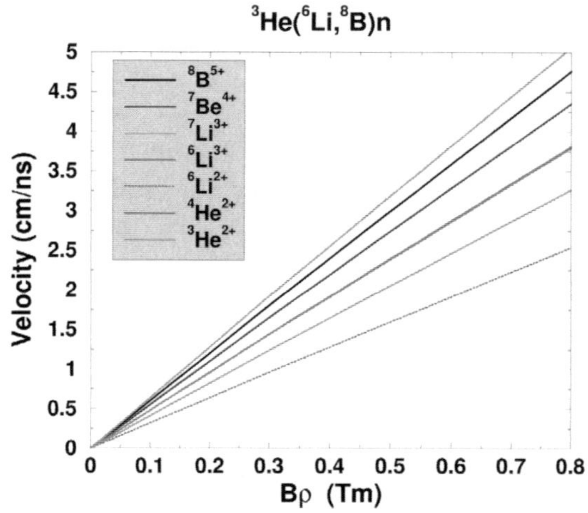

Figure 11. ^{8}B RIB production: kinetic energy as a function of the magnetic rigidity for the following beams: ^{8}B^{5+} (black line), ^{7}Be^{4+} (red line), ^{7}Li^{3+} (green line), ^{6}Li$^{3+,2+}$ (red lines), ^{4}He^{2+} (pink line) and ^{3}He^{2+} (yellow line).

Definitely more puzzling is the production of a ^{8}B RIB. In fact, besides the fact that the ^{8}B^{5+} RIB is by far the most energetic beam (see Fig. 10) which can be produced by means of the reaction ^{6}Li + ^{3}He, its velocity is exactly in between those of the ^{7}Be^{4+} contaminant beam and of fully striped ^{3}He ions recoiling out from the gas target. In this case, depending on the fine tuning of the WF, one can let either more ^{3}He^{2+} or more ^{7}Be^{4+} pass through the facility and reach the secondary target. The use of high transparency tracking detectors upstream the final focal plane of the facility may help disentangling the cocktail beam, since the particle velocity is selectively perturbed by the atomic interaction with the detector foils according to the atomic number of the ion species. Fig. 1 of Ref. [6] shows the actual situation occurred during an experiment aimed at measuring the fusion cross section for the system ^{8}B + ^{28}Si at Coulomb barrier energies. Depending on the beam energy and WF tuning, secondary beam purities in the range 30-43% were achieved.

In summary, the facility EXOTIC has already delivered seven RIBs in the energy range 1.5-6 MeV/u: ^{7}Be, ^{8}Li, ^{11}C, ^{17}F (with intensities larger than 10^5 pps), ^{15}O (intensity greater than 10^4 pps), ^{10}C and ^{8}B (intensities around 10^3 pps).

For the sake of completeness, we underline that there exist other five facilities in the world which produce and deliver RIBs in a similar manner: CRIB (Japan) [7, 8], MARS (USA) [9], TWINSOL (USA) [10], ANL (USA) [11] and RIBRAS (Brazil) [12].

5. Perspectives

The facility EXOTIC has been operational for more than a decade for the production of RIBs by means of the IF technique. A first experiment of astrophysical interest has been recently performed [13]. The goal of this experiment was the measurement of the cross section for the reaction ^{7}Be + $n \rightarrow \alpha + \alpha$ in the energy range up to E_{cm} = 1 MeV by means of the Trojan Horse Method [14] and the surrogate reaction ^{7}Be + ^{2}H $\rightarrow \alpha + \alpha$ + p. This study is particularly relevant for the cosmological ^{7}Li problem [15], since this process represents about 2.5% of the ^{7}Be destruction rate in the standard Big-Bang Nucleosynthesis model. The experiment was performed in November 2015 and a remarkable secondary beam intensity of 1.3×10^6 pps was achieved throughout the 13 days of beam time. The data analysis has just started.

Acknowledgements

This work was partially supported by the Italian Minister for Education, University and Research (MIUR) within the project RBFR08P1W2_001 (FIRB 2008).

References

[1] Smith M S and Rehm K E 2001 *Ann. Rev. Nucl. Part. Sci.* **51** 91.
[2] Farinon F et al. 2008 *Nucl. Instrum. Meth.* B **266** 4097.
[3] Mazzocco M et al. 2013 *Nucl. Instrum. Meth.* B **317** 223.
[4] Mazzocco M et al. 2010 *Phys. Rev.* C **82** 054604.
[5] Mazzocco M et al. 2015 *Phys. Rev.* C **92** 024615.
[6] Pakou A et al. 2013 *Phys. Rev.* C **87** 014619.
[7] Yanagisawa Y et al. 2005 *Nucl. Instrum. Meth.* A **539** 74.
[8] Yamaguchi H et al. 2008 *Nucl. Instrum. Meth.* A **589** 150.
[9] Tribble R E et al. 1989 *Nucl. Instrum. Meth.* A **289** 441.
[10] Becchetti F D et al. 2003 *Nucl. Instrum. Meth.* A **505** 377.
[11] Harss B et al. 2000 *Rev. Sci. Instrum.* **71** 380.
[12] Lichtenhaler R et al. 2005 *Eur. Phys. J.* A **25** s733.
[13] Lamia L et al. 2015 Proposal INFN-LNL 15.13.
[14] Tribble R E et al. 2014 *Rep. Prog. Phys.* **77** 106901.
[15] Broggini C et al. 2012 *Jour. Cosm. Astrop. Phys.* **6** 30.

Journal of Physics: Conference Series **703** (2016) 012013 doi:10.1088/1742-6596/703/1/012013

Application of few-nucleon physics in astrophysics

Michele Viviani[a] **and Laura Elisa Marcucci**[a,b]

[a] INFN, Sezione di Pisa, Pisa (Italy)

[b] Department of Physics, University of Pisa

E-mail: michele.viviani@pi.infn.it

Abstract. In this contribution a brief overview of the status and recent developments of *ab-initio* studies of nuclear reactions of astrophysical interest is presented.

1. Introduction

The study of nuclear reactions occurring in the cosmos are one of the primary goal of nuclear physics. These reactions enter in a large variety of phenomena, ranging from Big Bang Nucleosinthesis (BBN) [1], the formation and evolution of stars [2], the structure of neutron stars [3], etc. In particular, the theoretical study of these reactions has been vigorously pursued by many years. The "standard" starting point is to consider the nucleus as composed of nucleons interacting via non-relativistic potentials mediated by pions and other mesons. Also with this "simplification", there are still several issues not completely understood, in spite of decades of studies. Since the reactions occurring in the cosmos are driven from strong, electro-magnetic (EM), and weak forces, it is necessary, in order to obtain accurate estimates of the reaction rates, to take care of the following issues: *i)* it is necessary to well understand the nature of the strong nucleon-nucleon (NN) and of three-nucleon (3N) interactions; *ii)* a consistent theory for the interaction of nucleons with EM and weak probes (nuclear EM and weak currents) is also required; *iii)* we need to develop reliable and sophisticated methods for taking into account the nuclear structure of the nuclei entering the reactions, and their initial/final state interactions.

Usually, these issues are connected between each other. For example, EM current conservation imposes a strict relation between the Hamiltonian H describing the strong interaction among the nucleons and the nuclear EM current operators, entering in the matrix elements of the EM transition. Analogous relations exist for the weak interaction [4, 5].

Traditionally in nuclear physics the nuclear forces and currents have been constructed adopting some particular meson exchange model, usually limited to consider only one-boson exchange contributions. This picture led to a generation of realistic NN interactions which describe the NN data set with high precision [6, 7, 8]. Also 3N potentials and the EM and weak current operators have been constructed from these meson exchange models [4], but the issue of the consistency between all these quantities still remains to be clarified. Moreover, it is difficult to assess a "theoretical uncertainties" of the results obtained employing these models, as there is not a systematic criterion to select the contributions to be taken into account.

More recently, the advent of effective field theoretical methods and chiral perturbation theory (χPT) started to provide a more solid basis for a construction of consistent nuclear forces and currents [9, 10]. In the next Section, a brief overview of such a development is presented. In

Section 3, the preliminary results of a preliminary application of this method to the calculation of a reaction rate of astrophysical interest, namely the reaction $p + d \to {}^3\text{He} + \gamma$, is reported. Finally, in the last section, a brief conclusion is given.

2. *Ab-initio* studies of nuclear reactions of astrophysical interest

Many reactions of astrophysical interest are driven by the EM and weak interactions, as for example the reactions $p+d \to {}^3\text{He}+\gamma$ or $p+p \to d+e^+ +\nu_e$. The transition rate is usually well calculated using first order perturbation theory with respect to the "perturbation" Hamiltonian H_I, the part of the Hamiltonian describing the interaction of nucleons with EM and weak external probes. Consequently, it is possible to reduce the problem to the calculation of transition matrix elements of the type $\langle\Psi_f, X_f|H_I|\Psi_i, X_i\rangle$, where $|\Psi_i\rangle$ and $|\Psi_f\rangle$ are the initial and final nuclear states, while $|X_i\rangle$ and $|X_f\rangle$ are the initial and final states of eventual photons and/or leptons appearing in the reaction. The states $|\Psi_i\rangle$ and $|\Psi_f\rangle$ have to be calculated as solutions of $H|\Psi_x\rangle = E_x|\Psi_x\rangle$, where $x = i,\ f$ (E_i and E_f are energies of the initial and final nuclear states, respectively), namely by fully taking into account the dynamics due to the strong interaction between the nucleons and the distortion induced by the long-range Coulomb interaction between the protons. The latter effect is clearly of paramount importance for studying reactions of astrophysical interest, which usually take place at very low energies.

The "perturbation" H_I can be written in general of the form $H_I = \int d^3x \mathcal{F}_\mu(\boldsymbol{x})\mathcal{J}^\mu(\boldsymbol{x})$, where $\mathcal{J}^\mu \equiv (\rho, \boldsymbol{j})$ is the so-called nuclear current, consisting of operators acting on the nuclear states. The terms $\mathcal{F}_\mu(\boldsymbol{x})$ are field-theoretic operators describing the emission/absorption of photons, leptons, etc. This latter part is usually well known, as these operators are derived directly from the Standard Model (SM) EM and weak interactions. On the other hand, the operators $\mathcal{J}^\mu(\boldsymbol{x})$ have a complicated form due to the effects of the strong force.

There are also reactions not involving the effects of the EM and weak interaction, but driven directly by the strong force and/or the distortion due to the Coulomb interaction. The *ab-initio* study of such reactions requires the direct solution of the Schroedinger equation $H\Psi = E\Psi$ for positive energies, taking into account the dynamics of all nucleons. For example, the reaction ${}^3\text{He} + {}^3\text{He} \to {}^4\text{He} + 2p$ involves the dynamics of six nucleons, and the transition amplitude can be calculated solving the Schroedinger equation with appropriate boundary conditions.

In this contribution, we will concentrate on the reactions driven by either the EM or the weak interaction. In the next subsection, a brief introduction to the χPT for the construction of H will be reported. In Subsect. 2.2 we extend the discussion to the nuclear EM and weak interactions entering H_I. Finally, in the last subsection, some of the most accurate methods used to solve the Schroedinger equation for a system of A nucleons will be briefly recalled.

2.1. Strong forces

Traditionally in nuclear physics the Hamiltonian has been taken in a nonrelativistic form in which pairwise interactions between nucleons are supplemented by 3N forces for systems with more than two nucleons.

As discussed before, in recent years, the problem of the determination of these interactions has been tackled using an effective field theory (EFT) approach, based on the so-called chiral symmetry. This symmetry is an approximate symmetry of Quantum Chromodynamics (QCD), the fundamental theory that describes the interactions of quarks and gluons—the symmetry becomes exact in the limit of vanishing quark masses. At the nuclear level, where the nucleons and pions are the relevant degrees of freedom, this symmetry plays a main role in constraining the forces between nucleons, as well as the interaction of the nucleons with other SM particles as leptons, photons, etc.

In this approach, one starts from the most general Lagrangian describing pions and nucleons which takes into account the (approximate) chiral symmetry of QCD. In particular, chiral symmetry requires that the pion couple to these baryons, as well as to other pions, by powers of its momentum Q and, as a consequence, the Lagrangian describing their interactions can be expanded in powers of Q/Λ_χ, where $\Lambda_\chi \sim 1$ GeV is the chiral symmetry breaking scale. Classes of Lagrangians emerge, each characterized by a given power of Q/Λ_χ, or equivalently a given order in the derivatives of the pion field and/or pion mass factors, and each containing a certain number of unknown parameters, the so called low-energy constants (LECs), which, in practice, are fixed by comparison with experimental data. The central point of this approach is the definition of a power counting scheme which permits a systematic truncation of the possible contributions to the interaction (the so-called χPT).

The NN potentials most widely used in the literature are those of Refs. [11] and [12], and they are developed to the next-to-next-to-next-to-leading-order (N^{3}LO) in the chiral expansion. Very recently, the potential at one additional order (N^{4}LO) has been constructed [13, 14]. The convergence of this chiral expansion, at least for the NN system, seems to be now well established, as follows by comparing the predictions of the N^{3}LO and N^{4}LO NN forces [13].

In order to be used in the Schroedinger equation these potentials have to be regularized for large relative momenta k (or conversely for small interparticle distances). This is usually performed by multiplying them by a cutoff function, depending on a cutoff parameter Λ, so that for $k \gg \Lambda$ the potentials vanish. This procedure is somewhat arbitrary, and each group has chosen different method of regularization. The dependence of the results on Λ has been often used to establish the "theoretical uncertainty" of the calculation. In fact, it is expected that, at convergence of the expansion over Q/Λ_χ, the results should be independent by the high-energy behavior of the potential, namely by the precise value adopted for the cutoff parameter Λ. Therefore, by varying Λ (within a reasonable range of values), one should have information about the accuracy of the theoretical calculation, in particular that one due to the truncation of the χPT series. To be noticed that more accurate methods of extracting the theoretical uncertainty of the calculations based on χPT have been recently proposed [13, 15].

In their current form, these potentials contains 27 parameters (coming from the unknown LECs). These parameters have been fitted to the NN database, with a χ^2 very close to 1. Therefore, these chiral potentials have reached an accuracy comparable to those achieved for the traditional NN potentials.

New mechanisms for the 3N force have also been proposed using the chiral approach. The 3N force appears at N^{2}LO and contains two additional LECs [16, 17, 18]. The development of a 3N force at N^{3}LO and N^{4}LO is currently under way [19].

2.2. Nuclear currents

Also the development of the theory of the nuclear current operators $\mathcal{J}^\mu$ was the object of extensive studies. In general, they can be written as sums of one-, two-, and many-body terms that operate on the nucleon degrees of freedom, namely

$$\rho(\boldsymbol{q}) = \sum_i \rho_i(\boldsymbol{q}) + \sum_{i<j} \rho_{ij}(\boldsymbol{q}) + \dots\ , \quad \boldsymbol{j}(\boldsymbol{q}) = \sum_i \boldsymbol{j}_i(\boldsymbol{q}) + \sum_{i<j} \boldsymbol{j}_{ij}(\boldsymbol{q}) + \dots\ . \tag{1}$$

Above $\boldsymbol{q}$ is the three-momentum transfer to the nuclear system in the transition.

As stated before, the EM current operator must satisfy the current conservation relation (CCR) $\boldsymbol{q} \cdot \boldsymbol{j}_{EM}(\boldsymbol{q}) = [H, \rho_{EM}(\boldsymbol{q})]$. Therefore, the CCR imposes rather strict constraints on these EM nuclear currents. The weak current, on the other hand, can be divided in a vector and an axial component. The vector part, from the vector current conservation hypothesis, is directly related to the isovector part of the EM current, and therefore it is in practice constrained by the same CCR given above. The axial current verifies a similar relation in the limit of vanishing pion mass (partially conserved axial current hypothesis), and this also imposes strict constraints on the form of this part [5].

In recent years, the development of the nuclear currents has been brought forth systematically using the EFT approach, both for the EM current [20, 21, 22] and the weak current [23, 24]. In this approach, the EM and weak currents can be obtained directly starting from the same Lagrangian used to construct the nuclear forces. The currents constructed in this way verify the CCR automatically order by order in the χPT with the corresponding nuclear forces. Also in the construction of the nuclear currents one has to apply cutoff functions in order to regularize them for large relative momenta. An application of the EFT currents to an EM reaction of astrophysical interest is described in Section 3.

2.3. Calculation of the wave functions

Another important aspect for the determination of a reaction rate is the computation of numerically accurate nuclear wave functions. They are obtained as solutions of the associated quantum mechanical problem

$$H\Psi(1,2,\ldots,A) = E\Psi(1,2,\ldots,A) \ , \tag{2}$$

where H is a nuclear Hamiltonian. In this section, we will limit ourselves to mainly recalling some of the most used techniques. In the Faddeev Equation (FE) approach [25, 26, 27, 28], Eq. (2) is transformed to a set of coupled equations for the Faddeev amplitudes, which are then solved directly (in momentum or coordinate space) after a partial wave expansion.

The Urbana-Argonne group has in recent years developed [29, 30] the Green's Function Monte Carlo (GFMC) method, which has been successfully applied to calculate the spectra of nuclei up to $A = 12$. This techniques consists in the computation of $\exp(-\tau H)\Phi(1,2,\ldots,A)$, where $\Phi(1,2,\ldots,A)$ is a trial wave function, using a stochastic procedure to obtain, in the limit of large τ, the exact ground state wave function Ψ [29, 30].

The Stochastic Variational Method (SVM) [31] and the Coupled Rearrangement Channel Gaussian method (CRCG) [32] provide a variational solution of Eq. (2) by expanding the (radial part of the) wave function in Gaussians. Another widely applied variational method is the expansion of the wave function in terms of the Hyperspherical Harmonic (HH) basis [33].

In the no-core shell model (NCSM) method [34, 35] the calculations are performed using a (translationally-invariant) harmonic-oscillator finite basis P and introducing an effective P-dependent Hamiltonian H_P to replace H in Eq. (2). The operator H_P is constructed so that the solution of the equation $H_P\Psi(P) = E_P\Psi(P)$ provides eigenvalues which quickly converge to the exact ones as P is enlarged. The effective interaction hyperspherical harmonic (EIHH) method [36, 37] is based on a similar idea, but the finite basis P is constructed in terms of the HH functions.

In many reaction of astrophysical interest, either the initial or the final wave function corresponds to a scattering state. So far, only the FE and HH methods have been extended to find accurate solutions of Schroedinger equation in the continuum [25, 33]. Recently, also the NCSM has been generalized to treat scattering states using the resonating group method [38].

3. Proton radiative capture on deuterons

The radiative capture $p+d \to {}^3\mathrm{He}+\gamma$ is a relevant process in many astrophysical environments. For instance, it is the second step in the chain of nuclear reactions which, starting from the proton-proton weak capture, allows to stars like our Sun to shine via *pp*-chain. Interest in this reaction is also present in the context of BBN [1], since it is one of the main processes through which deuterium can be destroyed and thus affects its eventual yield.

The calculation reported in this Section has been performed as follows. The nuclear Hamiltonian H includes the non-relativistic kinetic energy, the chiral NN potential at N^{3}LO of Ref. [12], the chiral 3N potential at N^{2}LO [18], and the point-Coulomb interaction between the protons.

The "perturbation" Hamiltonian describing the EM interaction of nucleons with photons has been constructed using the chiral technique as well, as described in Ref. [20]. Although both operators in H and H_I have been determined using the same χPT technique, some details of the construction are not consistent. In particular, the chiral order of the NN force and the EM currents is not consistent, the EM currents do not include a three body term (necessary to verify the CCR with the 3N force), and the cutoff functions used for the NN force and the current are not the same. Consequently, the EM nuclear currents used do not exactly verify CCR. Work is in progress to solve this problem.

The present calculation has been performed by using two values of the cutoff parameter Λ (= 500 and 600 MeV) in order to have a hint of the theoretical uncertainty related to the truncation of the chiral expansion.

The initial and final wave functions have been calculated using the HH method, as described in detail in Refs. [33, 39]. The initial wave function is that one of the $p + d$ scattering, while the final wave function is that of the ^{3}He bound state. The theoretical uncertainties arising from the solution of the corresponding Schoredinger equations are very small and can be safely neglected [39].

The preliminary astrophysical S-factor obtained in this way has been reported in Fig. 1, where it is compared with the recent calculation of Ref. [40] performed using the traditional approach, as well as with the existing data of Refs. [41, 42, 44, 43]. The results are presented as a band corresponding to the two selected values of cutoff parameter Λ. The width of the band is rather small, suggesting that the theoretical error due to the truncation of the χPT is tiny for this process.

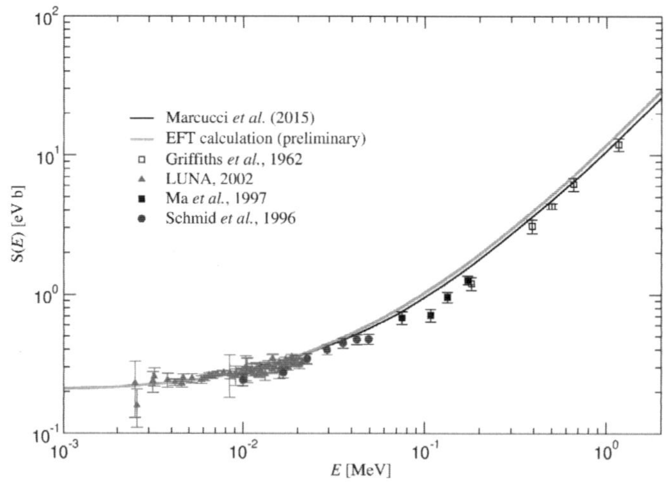

Figure 1. (Color online) Astrophysical S-factor of the reaction $p + d \to {}^3\text{He} + \gamma$, plotted as function of the center-of-mass kinetic energy E of the $p+d$ initial state. The preliminary results obtained using χPT as described in the text are given as the orange band. These results are compared with the available experimental data of Refs. [41, 42, 44, 43] and the calculation of Ref. [40] (solid black line).

The present results are in good agreement with the previous calculation and the LUNA data at low energies. Above $E = 50$ keV, the calculated theoretical S-factor is slightly larger than the available experimental data and that of Ref. [40]. Work is in progress to upgrade this χPT calculation.

Journal of Physics: Conference Series **703** (2016) 012013 doi:10.1088/1742-6596/703/1/012013

4. Conclusions

In recent years, there have been important developments in the construction of the nuclear forces and currents, and in the implementations of techniques for accurately solving (at least for few-nucleon systems) the Schroedinger equation even in presence of complicate NN and 3N force components. In particular, the advent of EFT method, strongly rooted on QCD, now provides a solid basis for the construction of consistent nuclear forces and currents. All these advances should in a near future allow for accurate studies of many reactions of astrophysical interest. Here we have presented only a preliminary study of one of them.

References

[1] See, for example, F. Iocco, *et al.*, Phys. Rep. **472**, 1 (2009) and references therein
[2] See, for example, E.G. Adelberger *et et al.*, Rev. Mod. Phys. **83**, 195 (2011), and references therein
[3] See, for example, J. M. Lattimer, Ann. Rev. Nucl. Part. Sci. **62**, 485 (2012) and references therein
[4] J. Carlson and R. Schiavilla, Rev. Mod. Phys. **70**, 743 (1998)
[5] J.D. Walecka, *Theoretical Nuclear and Subnuclear Physics*, Oxford University Press, 1995
[6] V.G.J. Stoks *et al.*, Phys. Rev. **C49**, 1950 (1994)
[7] R.B. Wiringa, V.G.J. Stoks, and R. Schiavilla, Phys. Rev. **C51**, 38 (1995)
[8] R. Machleidt, Phys. Rev. C **63**, 024001 (2001)
[9] S. Weinberg, Physica A **96**, 327 (1979); Phys. Lett. B **251**, 288 (1990); Nucl. Phys. B **363**, 3 (1991); Phys. Lett. B **295**, 114 (1992)
[10] C. Ordonez, L. Ray, and U. van Kolck, Phys. Rev. C **53**, 2086 (1996)
[11] E. Epelbaum, W. Gloeckle, and U.-G. Meissner, Nucl. Phys. A **637**, 107 (1998); *ibid.*, **747**, 362 (2005)
[12] D.R. Entem and R. Machleidt, Phys. Rev. C **68**, 041001 (2003)
[13] E. Epelbaum, H. Krebs, and U.-G. Meissner, Phys. Rev. Lett. **115**, 122301 (2015);
[14] D.R. Entem *et al.*, Phys. Rev. C **91**, 014002 (2015); *ibid.* **92**, 064001 (2015)
[15] H. W. Griesshammer, `ArXiv:1511.00490`
[16] C. Ordonez and U. van Kolck, Phys. Lett. **B291**, 459 (1992)
[17] U. van Kolck, Phys. Rev. C **49**, 2932 (1999)
[18] E. Epelbaum *et al.*, Phys. Rev. C **66**, 064001 (2002)
[19] V. Bernard *et al*, Phys. Rev. C **77**, 064004 (2008); H. Krebs, A. Gasparyan, and E. Epelbaum, Phys. Rev. C **85**, 054006 (2012)
[20] S. Pastore, *et al.*, Phys. Rev. C **80**, 034004 (2009)
[21] S. Kölling, *et al.*, Phys. Rev. C **80**, 045502 (2009); *ibid.* **84**, 054008 (2011)
[22] M. Piarulli, *et al.*, Phys. Rev. C **87**, 014006 (2013)
[23] T.-S. Park, D.-P. Min, and M. Rho, Phys. Rep. **233**, 341 (1993); Nucl. Phys. A **596**, 515 (1996)
[24] A. Baroni *et al.*, `[arXiv:1509.07039]`
[25] W. Glöckle *et al.*, Phys. Rep. **274**, 107 (1996)
[26] N. W. Schellingerhout, J.J. Schut and L. P. Kok, Phys. Rev. C **46**, 1192 (1992)
[27] A. Nogga *et al.*, Phys. Rev. C **65**, 054003 (2002)
[28] R. Lazauskas and J. Carbonell, Phys. Rev. C **70**, 044002 (2004)
[29] B.S. Pudliner *et al.*, Phys. Rev. C **56**, 1720 (1997)
[30] J. Carlson *et al.*, Rev. Mod. Phys. **87**, 1067 (2015)
[31] K. Varga and Y. Suzuki, Phys. Rev. C **52**, 2885 (1995)
[32] H. Kameyama, M. Kamimura, and Y. Fukushima, Phys. Rev. C **40**, 974 (1989)
[33] A. Kievsky *et al.*, J. Phys. G **35**, 063101 (2008)
[34] P. Navrátil, J. P. Vary, and B.R. Barrett, Phys. Rev. Lett. **84**, 5728 (2000); Phys. Rev. C **62**, 054311 (2000)
[35] P. Navrátil and W.E. Ormand, Phys. Rev. Lett. **88**, 152502 (2002)
[36] N. Barnea, W. Leidemann, G. Orlandini, Phys. Rev. C **61**, 054001 (2000)
[37] N. Barnea, W. Leidemann, and G. Orlandini, Phys. Rev. C **67**, 054003 (2003)
[38] G. Hupin, S. Quaglioni, and P. Navrátil, Phys. Rev. C **90**, 061601 (2014)
[39] L.E. Marcucci *et al.*, Phys. Rev. C **80**, 034003 (2009)
[40] L.E. Marcucci *et al.*, `[arXiv:1510.07877]`
[41] G.M. Griffiths, M. Lal, C.D. Scarfe, Can. J. Phys. **41**, 724 (1962)
[42] G.J. Schmid *et al.*, Phys. Rev. Lett. **76**, 3088 (1996)
[43] C. Casella *et al.*, (LUNA Collaboration), Nucl. Phys. A **706**, 203 (2002)
[44] L. Ma *et al.*, Phys. Rev. C **55**, 588 (1997)

Journal of Physics: Conference Series **703** (2016) 012014 doi:10.1088/1742-6596/703/1/012014

Status of the direct measurements of $^{18}O(p,\gamma)^{19}F$ and $^{23}Na(p,\gamma)^{24}Mg$ cross sections at astrophysical energies at LUNA

A Boeltzig[1], F R Pantaleo[2,3], A Best[4], G Imbriani[5,6] and M Junker[1,4]

[1] Gran Sasso Science Institute, L'Aquila, Italy
[2] Università degli Studi di Bari Aldo Moro, Bari, Italy
[3] INFN Sezione di Bari, Bari, Italy
[4] Laboratori Nazionali del Gran Sasso, Assergi, Italy
[5] Università degli Studi di Napoli Federico II, Napoli, Italy
[6] INFN Sezione di Napoli, Napoli, Italy

E-mail: axel.boeltzig@gssi.infn.it

Abstract. $^{18}O(p,\gamma)^{19}F$ and $^{23}Na(p,\gamma)^{24}Mg$ are reactions of astrophysical interest for example in AGB star scenarios. The rates of both reactions are potentially influenced by low-energy resonances for whose strengths either exist only values with large uncertainties, upper limits or even contradictory claims. Measurements at the Laboratory for Underground Nuclear Astrophysics (LUNA) aim at a direct observation of these low-energy resonances, and additional cross section measurements to aid a more precise determination of the reaction rates in astrophysical scenarios. We report the experimental setup and the status of the ongoing measurements of the two reactions at LUNA.

1. Motivation and Objectives

The $^{18}O(p,\gamma)^{19}F$ reaction ($Q \approx 7.99\,\mathrm{MeV}$) competes with $^{18}O(p,\alpha)^{15}N$ in AGB star nucleosynthesis and may provide an explanation for an observed ^{18}O depletion in presolar grains. In the temperature range up to 10^8 K, the rate of this reaction is determined by direct capture and up to three resonances at center of mass energies of 19 keV, 90 keV and 144 keV. So far, only the 144 keV resonance has been observed in direct measurements. A recent experiment resulted in an upper limit for the resonance strength of $\omega\gamma < 7.8\,\mathrm{neV}$ (90% CL) [1]. Using data obtained from $^{18}O(p,\alpha)^{15}N$ and $^{18}O(^3He, d)^{19}F$, a value for $\omega\gamma$ that is 90(36) times larger than this limit was derived in [2]. This resonance strength would enhance the reaction rate significantly for temperatures around 50 MK. To resolve this discrepancy, a closer study of this resonance in a direct measurement is thus desirable. Additionally, measurements of the non-resonant cross section below the 144 keV resonance with a low-background setup would complete available data from earlier measurements [3].

The reaction $^{23}Na(p,\gamma)^{24}Mg$ ($Q \approx 11.69\,\mathrm{MeV}$) links the NeNa and MgAl cycles, competing with $^{23}Na(p,\alpha)^{20}Ne$. The reaction rates up to temperatures of 10^9 K are determined by the direct capture component and three narrow resonances at 138 keV, 241 keV and 296 keV center of mass energy. A direct measurement of the strength of the 138 keV resonance [4] resulted in an upper limit of $\omega\gamma < 5.17\,\mathrm{neV}$ (95% CL) and an indication of a signal at $\omega\gamma = 2.15(129)\,\mathrm{neV}$

(68% CI). For the latter value, the 138 keV resonance would be the dominant contribution to the reaction rate for temperatures around 80 MK. A confirmation and more precise determination of the resonance strength would reduce the reaction rate uncertainty in this temperature range. Additionally the non-resonant cross section for this reaction is of interest.

The Laboratory for Underground Nuclear Astrophysics (LUNA) is dedicated to the direct measurement of cross sections at low energies. Thanks to its underground location at the Gran Sasso National Laboratory (LNGS) it benefits greatly from a reduced cosmic-ray background, which also allows the construction of massive shielding setups against environmental radioactivity. The low background levels that are achieved this way allow for the measurement of reactions with small cross sections. The efforts to study $^{18}O(p,\gamma)^{19}F$ and $^{23}Na(p,\gamma)^{24}Mg$ at LUNA use the same setup. They aim at the study of the resonances at 90 keV in $^{18}O(p,\gamma)^{19}F$ and 138 keV in $^{23}Na(p,\gamma)^{24}Mg$. Measurements on the 241 keV resonance in $^{23}Na(p,\gamma)^{24}Mg$ and the non-resonant contribution for both reactions are further objectives.

2. Setup

The LUNA accelerator provides a beam of protons with energies up to 400 keV. Beam currents of 100-200 μA on target are achieved. A solid target setup, installed on one of the two beamlines present at LUNA, is used for studies of the two reactions. Solid target setups have been used in previous measurements at LUNA, the setup has been refurbished and extended for the current measurements.

Tantalum sheets are used as backings for both target types. Oxygen targets are produced by anodization, a well-established technique already used at LUNA [5]. Sodium targets are obtained by evaporation of sodium salts on the backing material. Required properties for these targets are a regular and well-defined target density profile, stability of the target composition under proton bombardment and high-purity from contaminants to avoid beam-induced backgrounds. To achieve these properties, different salts are under study, including Na_2WO_4 that has been used in earlier studies such as [4]. The target density profiles are studied and monitored in situ using narrow resonances of well-known strength in $^{18}O(p,\gamma)^{19}F$ and $^{23}Na(p,\gamma)^{24}Mg$. Further analyses ex situ are employed in the development of Na targets.

For the resonance searches a segmented BGO detector with a solid angle coverage close to 4π [6] is used, as it provides a large full energy detection efficiency (on the order of 50% for a single 10 MeV gamma ray) at a moderate energy resolution (17% FWHM at 662 keV from ^{137}Cs).

The experimental setup also allows to use a High Purity Germanium (HPGe) detector instead of the BGO, with relatively small changes in the setup. This detector provides a much lower detection efficiency compared to the BGO detector but a greatly improved energy resolution. It is therefore well-suited to measure individual gamma ray intensities and deduce branching ratios at proton energies where the reaction rate is sufficiently large for detection given the efficiency of the detector.

3. Backgrounds

The environmental backgrounds in the BGO detector, in particular the γ ray component, are reduced by a 10 cm lead shielding surrounding the detector. With this setup the dominant environmental background in the signal region of interest for $^{18}O(p,\gamma)^{19}F$, i.e. around its Q value, is dominated by radiative capture of thermal neutrons in the detector materials, but only on the order of 10^{-4} counts per MeV per second. In the case of $^{23}Na(p,\gamma)^{24}Mg$ the Q value lies significantly above the largest Q value of radiative neutron capture, and hence the environmental background is even lower.

Beam-induced backgrounds from reactions on beamline elements or in the target can be problematic if the Q value or the cross section of these reactions is large. Examples for potentially critical background reactions from contaminants include $^{7}Li(p,\gamma)^{8}Be$, $^{11}B(p,\gamma)^{12}C$

Journal of Physics: Conference Series **703** (2016) 012014 doi:10.1088/1742-6596/703/1/012014

and ^{19}F(p,$\alpha\gamma$)^{16}O. A background due to the composition of the sodium tungstate targets is ^{18}O(p, γ)^{19}F, relevant in vicinity of the stronger resonances of this reaction.

The importance of these backgrounds depend on the proton energy and the reaction under study. The ^{11}B(p,γ)^{12}C resonance, for example, is rather critical for the 138 keV resonance search in ^{23}Na(p, γ)^{24}Mg, as it has a broad resonance at 161 keV proton energy and a peak in the detected total energy spectrum that is very close to the expected signal from the resonance in ^{23}Na(p, γ)^{24}Mg. The reduction of beam-induced background contributions by careful material selection, and the subtraction and discrimination of backgrounds during the analysis are important to achieve the necessary sensitivity for competitive measurements.

4. Measurements

Measurements for both reactions are ongoing with the BGO detector setup. The non-resonant component of ^{18}O(p, γ)^{19}F has been measured from slightly above the 90 keV resonance up to 400 keV, including measurements of the resonances in this energy range for the purpose of cross checks of the normalization. The focus of the current measurements is on the search for the 90 keV resonance, and possibly the measurement of the non-resonant cross section below this resonance.

The ^{23}Na(p, γ)^{24}Mg measurements started with a focus on the 138 keV resonance search, measuring the 296 keV resonance at regular intervals as a normalization. The uncertainty of the energy of the low energy resonance is 4.0 keV [7] or 1.5 keV [8], which requires a trade-off for the proton energy studied to be high enough to be above the resonance energy, but low enough to reduce backgrounds from the broad ^{11}B(p,γ)^{12}C resonance discussed above. As a compromise, a set of energies around the central value of the resonance energy was chosen for measurements. Additionally, the non-resonant cross section is being measured.

5. Conclusions and outlook

Measurements of the ^{18}O(p, γ)^{19}F and ^{23}Na(p, γ)^{24}Mg with a high-efficiency BGO detector are currently running at LUNA. For both reactions the current phase of measurements focuses on the search for known resonances at low energies and the measurement of non-resonant cross section contribution where possible.

After this phase, measurements of both reactions with a HPGe detector are foreseen. These measurements will focus on the determination of gamma ray branching ratios in the energy ranges where the reaction rate is sufficiently high for a measurement, given the lower efficiency of the HPGe compared to the BGO detector. In particular these measurements will concern the 241 keV resonance in ^{23}Na(p, γ)^{24}Mg and the non-resonant cross section in both reactions, where the energy ranges to be studied will be guided by the results obtained in the BGO detector phase.

References

[1] Buckner M Q *et al.* 2012 *Phys. Rev.* C **86** 065804
[2] Fortune H T 2013 *Phys. Rev.* C **88** 015801
[3] Wiescher M *et al.* 1980 *Nucl. Phys.* A **349** 165–216
[4] Cesaratto J M *et al.* 2013 *Phys. Rev.* C **88** 065806
[5] Caciolli A *et al.* 2012 *Eur. Phys. J.* A **48** 144
[6] Casella C *et al.* 2002 *Nucl. Instr. Meth. Phys. Res.* A **489** 160–169
[7] Firestone R B 2007 *Nucl. Data Sheets* **108** 2319–2392
[8] Hale S E *et al.* 2004 *Phys. Rev.* C **70** 045802

An overview of the ^{19}F(p,α_0)^{16}O reaction with direct methods

D. Dell'Aquila and I. Lombardo

Dipartimento di Fisica, Università di Napoli Federico II and INFN Sezione di Napoli, 80126 Napoli, Italy

E-mail: dellaquila@na.infn.it

Abstract. The study of the ^{19}F(p,α)^{16}O reaction at low energy is important both for Nuclear Structure and Astrophysics. Despite of its importance, the S-factor of this reaction is poorly known, especially at astrophysical energies. We present an overview of the ^{19}F(p,α_0)^{16}O reaction cross section, as obtained from recent direct measurements and from published works in the literature. We include in the systematic also data from an unpublished work, where several excitation functions and angular distributions for α_0 and α_π channels are reported.

1. Introduction

The ^{19}F(p,α)^{16}O reaction can provide important information both for the study of Nuclear Structure of ^{20}Ne [1] and the nucleosynthesis of Fluorine and heavier elements in massive stars [2]. In the first case it allows to investigate the spectroscopy of natural parity states in the compound ^{20}Ne nucleus [3] above the proton separation threshold (S_p = 12.844 MeV [4]), where the possible existence of ^{20}Ne quartet excitations has been predicted in various theoretical works [5, 6].

Furthermore, the direct investigation of the ^{19}F(p,α)^{16}O reaction at energies well lower then the Coulomb barrier is very useful for solving astrophysical problems related to the understanding of CNO-F cycles in hydrogen burning stages of stellar evolution [2] and the nucleosynthesis of fluorine [7]. In fact, this reaction represents the branching point between the CNO cycles and the NeNa one [2] and the amount of catalytic material that is lost from the CNO cycles to activate the NeNa one is regulated by the competition between the ^{19}F(p,α) and ^{19}F(p,γ) reactions [2, 8, 9]. Another interesting problem is connected to the galactic fluorine abundance. It is strongly dependent on the condition within the astrophysical sites and therefore it is used to probe different nucleosynthesis scenarios [7]. Moreover, asymptotic giant branch (AGB) stars are regarded as the major contributors to the fluorine abundance in the Milky Way [10], but the observed upper limit of fluorine abundance is well lower then the one predicted with the most recent AGB models [11]. This may be caused by possible deep mixing phenomena that could expose the catalytic material at temperature high enough to activate the fluorine destruction via the ^{19}F(p,α) reaction.

Despite its importance, very few direct investigations of the ^{19}F(p,α)^{16}O reaction at low energies have been reported in literature. These data would be useful both to directly determine the presence of low energy resonances and to give good reference points for the absolute normalization of data obtained with indirect methods, like the Trojan Horse (THM) one [12, 13].

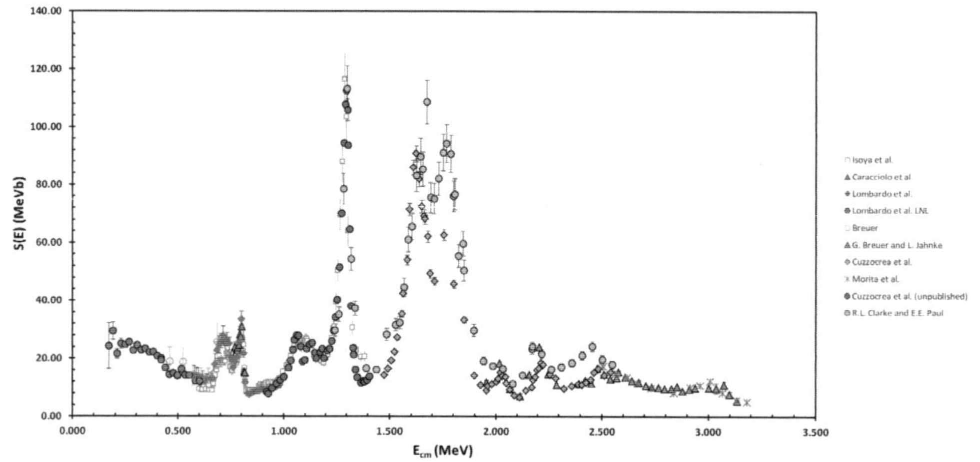

Figure 1. ^{19}F(p,α_0) S-factor data sets in the $0.2\text{MeV} \leq E_{cm} \leq 3.3\text{MeV}$ energy range. Red circles represent the unpublished data from Ref. [22], while the data sets from Refs.[3, 21, 23] (respectively blue circles, red rhombuses and blue triangles) are not included in the NACRE compilation.

In this proceeding we report an overview of the ^{19}F(p,α_0)^{16}O direct experimental data in a quite large energy range ($0.2\text{MeV} \leq E_{cm} \leq 3.3\text{MeV}$), to try to clarify some ambiguities still present in the S-factor.

2. Astrophysical S-factor: an overview

The integrated cross section and S-factor of the ^{19}F(p,α_0)^{16}O reaction have been the subject of direct experiments performed in past times. In particular, the Nuclear Astrophysics Compilation of Reaction Rates (NACRE [14]) reports the following direct data sets in the $E_{cm} = 0.4-3.2\text{MeV}$ energy domain: Morita et al. [15], Clarke and Paul [16], Cuzzocrea et al. [17], Isoya et al. [18], Caracciolo et al. [19] and Breuer [20].

The higher energy points, red stars in Fig. 1, come from the old paper of Morita et al. [15], where the energy domain $2.64\text{MeV} \leq E_p \leq 3.35\text{MeV}$ was covered. The corresponding integrated cross section was obtained in relative units and has been normalized by NACRE to 28mb at $E_{cm} \simeq 2.5\text{MeV}$ for comparison with the other data sets. These data overlap nicely with the absolute data from Breuer and Jahnke [21] (blue triangles), not included in the NACRE compilation. In the $1.5\text{MeV} \leq E_{cm} \leq 2.5\text{MeV}$ we observe the presence of differences between the relative data of Clarke and Paul [16] (yellow points) and the absolute data by Cuzzocrea et al. [17] (green triangles). In particular, the Clarke and Paul data, normalized by NACRE to $\sigma = 42\text{mb}$ at the $E_p \approx 1.3\text{MeV}$ resonance, seem to overrate the S-factor in their high energy part.

The data from Isoya et al, carried out by means of a proportional counter and a proton beam covering an energy range from 0.630MeV to 1.460MeV, have also been normalized by NACRE at the $E_p \approx 1.3\text{MeV}$ resonance. These experimental points are in good agreement with ones from Caracciolo et al. [19] (green triangles), in the region of the $E_{cm} = 0.842\text{MeV}$ ^{20}Ne resonance, but disagree at lower energies with Breuer [20] (light blue open circles), leading to strong uncertainties, of the order of 50% [14], in the S-factor extrapolation down to the Gamow peak.

The behaviour of the S-factor at lower energies could be better understood by considering the results of two recent experiments carried out in Naples (Lombardo et al. [3, 24, 25], $E_p \simeq$

$0.6-1$MeV) and at Laboratori Nazionali di Legnaro (Lombardo et al. [23], $E_p \simeq 0.2-0.6$MeV), indicated with blue circles in Fig. 1. These two experiments confirm the S-factor rise at low energies seen in Ref. [20], and point out the presence of broad resonances at low energies, in agreement with indirect observations made with the Trojan Horse Method [7, 26].

Finally, we included in the systematic of data also unpublished results, discussed in an INFN Activity Report, by Cuzzocrea et al. [22]. In this work, very fine step ($\approx$ 10keV) excitation functions are reported in the 0.9MeV $\leq E_p \leq$ 1.MeV domain for 8 different detection angles. By extracting the data and integrating the corresponding angular distributions we obtained the S-factor showed with the red circles in Fig. 1. This data set is consistent with the high energy part of the data from Lombardo et al. [3] and it is in quite nice agreement with the two relative data sets from Refs. [16, 18], confirming the consistency of the NACRE normalization at the $E_p \approx 1.3$ MeV resonance.

3. Conclusions and Perspectives

We discuss an overview of direct data sets of the ^{19}F(p,α_0)^{16}O reaction cross section in the energy range $E_{cm} \leq 3.3$MeV. We considered the data reported in the NACRE compilation and the data from Refs. [3, 21, 23], not indicated by NACRE. By including the unpublished results from Ref. [22] we are able to give absolute values of the S-factor in the $E_p \approx 1.3$MeV resonance region, where only relative data sets were reported in the literature. For the future, we plan to perform a R-matrix fit of this data set, together with other reaction and scattering channels populating ^{20}Ne excited states, to improve the spectroscopy of this self-conjugated nucleus.

4. Acknowledgments

This work has been done in collaboration with L. Campajola, P. Cuzzocrea, A. Di Leva, I. Indelicato, M. La Cognata, M. La Commara, A. Ordine, E. Perillo, V. Rigato, M. Romoli, E. Rosato, G. Spadaccini, C. Spitaleri, A. Tumino and M. Vigilante.

References

[1] Tilley D *et al.* 1998 *Nucl. Phys. A* **636** 249
[2] Rolfs C and Rodney W 1988 *Cauldrons in the Cosmos* W.S. (Chicago: Univ. Chicago Press)
[3] Lombardo I *et al.* 2013 *J. Phys. G: Nucl. Part. Phys.* **40** 125102
[4] NNDC database `http://www.nndc.bnl.gov/`
[5] Arima A, Gillet V and Ginocchio J 1970 *Phys. Rev. Lett.* **25** 1043
[6] Middleton R, Garret J and Fortune H 1971 *Phys. Rev. Lett.* **27** 950
[7] LaCognata M *et al.* 2011 *Astr. Jour. Lett.* **739** L54
[8] Couture A *et al.* 2008 *Phys. Rev. C* **77** 015802
[9] Spyrou K *et al.* 2000 *Eur. Phys. J. A* **7** 79
[10] Jorissen A *et al.* 1992 *A&A* **261** 164
[11] Lugaro M *et al.* 2004 *Astr. J.* **615** 934
[12] Baur G 1986 *Phys. Lett. B* **178** 135
[13] Spitaleri C *et al.* 1999 *Phys. Rev. C* **60** 055802
[14] Angulo C *et al.* 1999 *Nucl. Phys. A* **656** 3
[15] Morita S *et al.* 1986 *J. Phys. Soc. Japan* **21** 2435
[16] Clarke R and Paul E 1957 *Can. J. Phys.* **35** 155
[17] Cuzzocrea P *et al.* 1980 *Lett. Nuovo Cim.* **28** 515
[18] Isoya A, Ohmura H and Momota T 1959 *Nucl. Phys.* **7** 116
[19] Caracciolo R *et al.* 1974 *Lett. Nuovo Cim.* **11** 33
[20] Breuer G 1959 *Z. Phys.* **154** 339
[21] Breuer G and Jahnke L 1964 *Z. Naturf.* **19a** 471
[22] Cuzzocrea P *et al.* 1980 report INFN/Be-80/5 4 (unpublished)
[23] Lombardo I *et al.* 2015 *Phys. Lett. B* **748** 178
[24] Lombardo I *et al.* 2014 *J. Phys. Conf. Ser.* **569** 012068
[25] Lombardo I *et al.* 2014 *Bull. Rus. Acad. Sci. Phys.* **78** 1093
[26] LaCognata M *et al.* 2015 *Astr. Jour.* **805** 128

First evidences for ^{19}F(α, p)^{22}Ne at astrophysical energies

D'Agata G.[1,2], Spitaleri C.[1,2], Pizzone R.G. [1], Blagus S.[4], Figuera P.[1], Grassi L.[2], Guardo G.L.[1,2], Gulino M.[1,3], Hayakawa S.[5], Indelicato I.[1,2], Kshetri R.[6], La Cognata M.[1], Lamia L.[2], Lattuada M.[1,2] , Mijatović T.[4], Milin M.[4], Miljanic D.[4], Prepolec L.[4], Sergi M.L.[1], Skukan N.[4], Soic N.[4], Tokic V.[4], Tumino A.[1,3], Uroic M.[4]

[1] INFN-LNS, Via Santa Sofia, 62, 95123 Catania CT, IT
[2] Dipartimento di Fisica e Astronomia, Università degli Studi di Catania, Via Santa Sofia, 64, 95123 Catania CT, IT
[3] Kore University, Viale delle Olimpiadi, 1, 94100 Enna EN, IT
[4] Rudjer Boskovic Institut, Bijenika cesta, 54, 10000, Zagreb, HR
[5] RIKEN, CNS, 2-1 Hirosawa, Wako, Saitama 351-0198, Japan
[6] Department of Physics, Sidho-Kanho-Birsha University, Purulia - 723104, WB, India

E-mail: g.dagata@lns.infn.it

Abstract. ^{19}F experimental abundances is overestimated in respect to the theoretical one: it is therefore clear that further investigations are needed. We focused on the ^{19}F(α, p)^{22}Ne reaction, representing the main destruction channel in He-rich environments. The lowest energy at which this reaction has been studied with direct methods is $E_{C.M.} \approx 0.91$ MeV, while the Gamow region is between $0.39 \div 0.8$ MeV, far below the Coulomb barrier (3.8 MeV). For this reason, an experiment at Rudjer Boskovic Institut (Zagreb) was performed, applying the Trojan Horse Method. Following this method we selected the quasi-free contribution coming from ^{6}Li(^{19}F,p ^{22}Ne)^{2}H at E_{beam}=6 MeV at kinematically favourable angles, and the cross section at energies $0 < E_{C.M.} < 1.4$ MeV was extracted in arbitrary units, covering the astrophysical region of interest.

1. Astrophysical background

Fuson reaction inside stars are the main responsible for element production if $A < 60$. Heavier elements are not produced in this way, because of the Coulomb barrier between the interacting nuclei. If $A > 60$, other processes are activated, such as the neutron capture (s-process and r-process), that do not contribute to the stellar energy production, but are important for heavy elements nucleosynthesis. Their producton takes place inside Asimptotic Giant Branch (AGB) stars [1], in which the synthetized isotopes are brought on the surfice by the so-called *third dredge up*: in this phase the star is caracterized by a degenerete carbon-oxygen core, surrounded by a helium and hydrogen shells, separeted by a "thin" layer ($10^{-2} - 10^{-3} R_{\odot}$), called helium-intershell. At high enough temperatures ($T \approx 10^{8}$ K), the ^{14}N produced in the CNO cycle could bring to the formation of ^{19}F, using the production chain $^{14}N(\alpha, \gamma)^{18}F(\beta^{+}\nu)^{18}N(p, \gamma)^{19}F$. It is important to underline that AGB stars are the only confirmed sites of ^{19}F production so far. The abundance of ^{19}F is not well reproduced by the various astrophysical models, being lower than what is observed. It is therefore important to study how ^{19}F is destroyed, given that the production pattern is quite clear. In AGB stars it can be destroyed by $^{19}F(p,\,\alpha)^{16}O$ [2] and $^{19}F(\alpha,\,p)^{22}Ne$, with the latter having greater importance in He-rich environments. This is the key argument of the present work. At typical temperatures ($T = 8 \cdot 10^8$ K) for low-mass ($2 \div 4\, M_\odot$) AGB stars, the Gamow window of astrophysical interest is located between 390 and 800 keV, while direct measurements have reached the lowest energies are at $E_{C.M.} = 0.91$ MeV [3][4], making extrapolation at lower energies necessary, where the cross-section is exponentially small. If Coulomb barrier is considered, another problem arises: the Coulomb barrier is at $E_c \simeq 3.8$ MeV, far above the Gamow window. Keeping this in mind is easy to understand how using indirect methods could be useful. Among them, the Trojan Horse Method (THM) [5] [6] [7] [8] are needed for this kind of reaction. In this work in particular, the three-body reaction $^{19}F(^6Li,\,\alpha\,^{22}Ne)^2H$ is used to study the $^{19}F(\alpha,\,p)^{22}Ne$ two-body reaction.

2. Experimental set-up

When approaching to the THM is important to verify some conditions regarding the *Trojan Horse* nucleus (whether projectile or target) and the particles in the exit channel. For the first of the two, we have to be sure that:

- The TH nucleus must show a cluster-like structure
- The Binding energy of the constituting cluster must be small in comparison to the beam energy
- The momentum distribution of the cluster inside the TH nucleus must be known

Keeping this in mind, it is possible to fix the experimental set-up in a way that maximizes the THM contribution to the reaction. We therefore studied the reaction $^{19}F(\alpha,\,p)^{22}Ne$ starting from the three-body reaction $^{19}F(^6Li,\,\alpha\,^{22}Ne)^2H$ using the Trojan Horse Method, which allows to have indirect measurements of the two-body cross section, avoiding inconveniences brought by Coulomb barrier, if the condition above are satisfied. In this experiment the ^6Li projectile was used as THM nucleus, given that it can be described as composed by an α particle (participant) and a deuteron (spectator) with large probability.

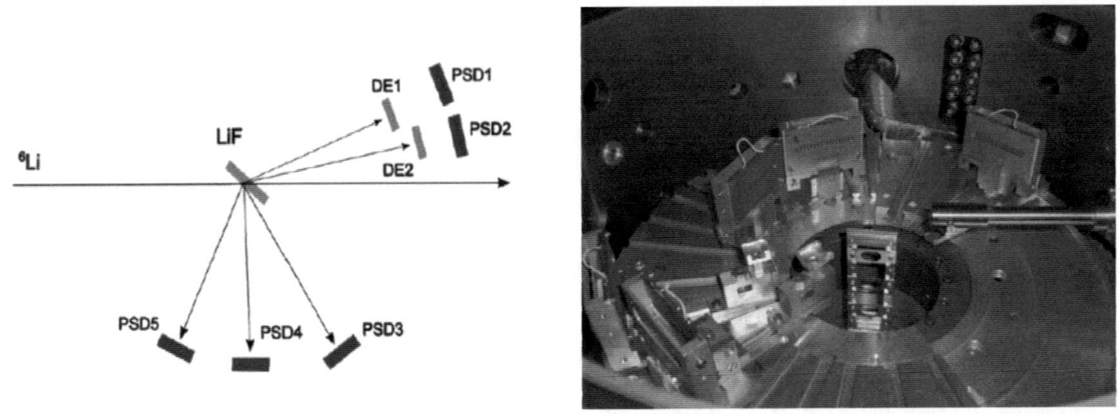

Figure 1: Experimental Set-up

Journal of Physics: Conference Series **703** (2016) 012016 doi:10.1088/1742-6596/703/1/012016

The experimental set-up is shown in Fig.1. The aim was to detect deuterium (spectator particle) and protons, coming from the three-body reaction. A beam of ^6Li hit a ^7LiF target, oriented at 45° with respect to the beam axis to reduce, on the average, the straggling of the particles emerging from the target. Following the THM prescriptions, beam energy was chosen to be 6 MeV. The detection apparatus was composed by two ΔE-E telescopes, made by a thick silicon detector (500 μm) and a thin one (9 : μm), placed at 12.3° and 32.3° meant for deuterium detection, and three other thick silicon detectors meant for proton detection (−37.3°, −81° and −119.9°). The events were registered by the acquisition system only if coincidence occurs between one of the detectors placed at one side of the beam and one on the other side. All the used detectors are position sensitive (PSD), because good angular resolution is crucial for such measurements. In this work we focused on coincidences between PSD2 and PSD3 (with respect to the Fig.1), given that this coincidence were proven to be the most favourable one.

3. Channel selection

By means of the standard ΔE-E technique we were able to identify incoming particles. Using them we were able to select deuterons, but we did not know which were effectively coming from the reaction of interest. In order to ascertain that, we needed to compare the expected theoretical Q-value with the experimental one. In Fig.2 is shown that the Q-value spectrum had a single peak centred at the theoretical value ($Q = 0.199$ MeV).Experimental data were also consistent with the simulated kinematic locus for the three-body reaction.

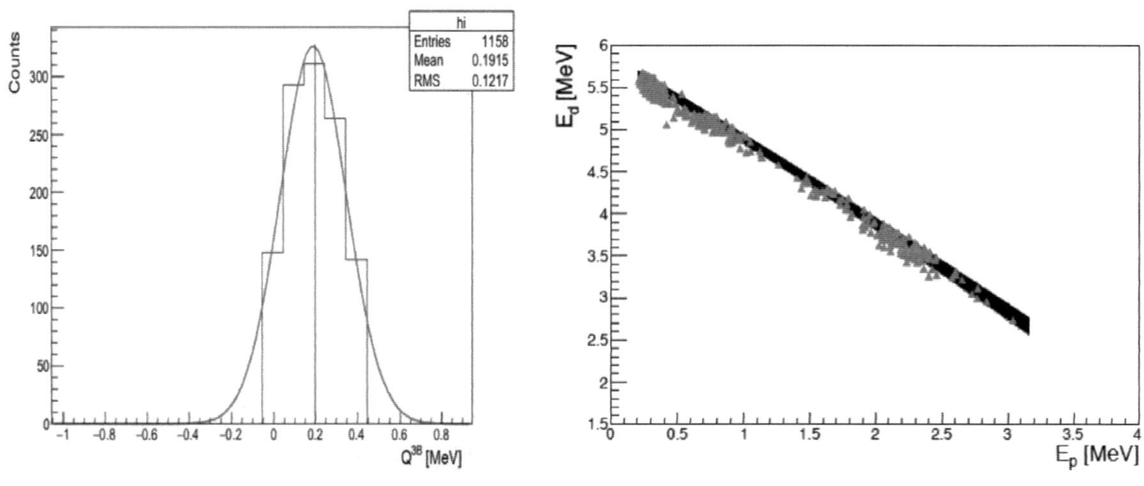

Figure 2: Left panel: experimental Q-value for the three body reaction (the blue line represent the theoretical value)
Right panel: comparison between $E_p vs. E_d$ experimental scatter plot and Monte Carlo simulation kinematic locus

The next step of the data analysis was to check if the reaction mechanism was predominantly a quasi-free reaction. For this purpose the experimental momentum distribution of the deuterium inside ^6Li, fitted with an Hankel function whose width is given by $W(q) = f_0(1 - e^{-\frac{q_t}{q_0}})$, with $q_t = p_{beam} - \frac{p_p + p_{22Ne}}{2}$ transferred momentum, f_0 asymptotical width of the function and $q_0 = 122 \pm 3.5$ MeV fit parameter, was compared with literature [9] (Fig3).

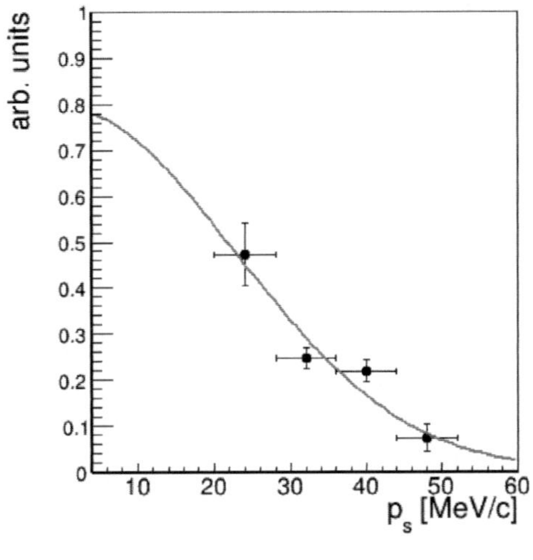

Figure 3: Deuterium momentum distribution inside ^6Li fitted with an Hankel function

4. Results

An estimation of the cross section was possible. Using the standard formulation for the differential cross-section in the THM, $\frac{d^3\sigma}{d\Omega_p d\Omega_{^{22}Ne} dE_{Ecc}} \propto \left[KF|\Phi(p_s)|^2\right] \times \left(\frac{d\sigma}{d\Omega}\right)^{HOES}$ [7], we obtained what is shown in Fig.4. A first estimation of the cross section (even if still in arbitrary units) for the reaction $^{19}F(\alpha,\ p)^{22}Ne$ at astrophysical energies was performed.

In Fig.4 the experimental results were fitted with a sum of several gaussian functions, whose centroids are consistent with the literature [10] (reported in Tab.1).

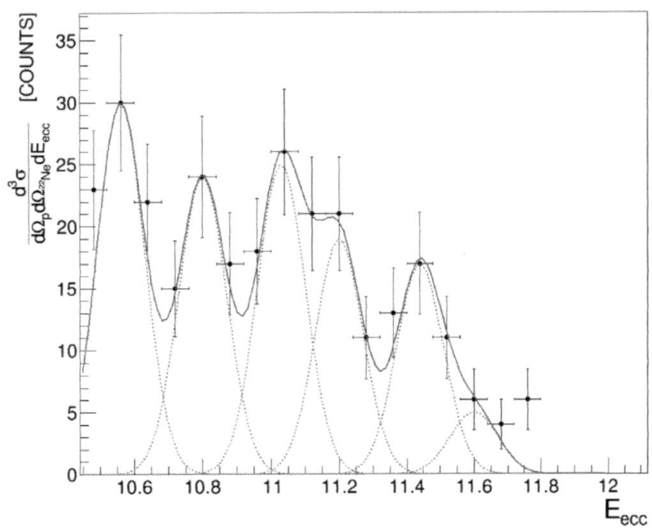

Figure 4: Triple differential cross section for the reaction of intrest, fitted using several gaussian functions)

E*($^{23}Na)^{th}$ [MeV]	J$^{\pi}$	E*($^{23}Na)^{exp}$ [MeV]	E$^{th}_{C.M.}$ [MeV]	E$^{exp}_{C.M.}$ [MeV]
10.575	$3/2^-$	10.6	0.107	0.1
10.823	$3/2^+$	10.8	0.356	0.3
11.038	$1/2^+$	11.0	0.571	0.6
11.238	$3/2^-$	11.2	0.771	0.8
11.355	$1/2^+$	11.4	0.887	0.9
11.554	$1/2^+$	11.6	1.086	1.1

Table 1: Experimentally desumed energy levels(as in Fig. 4)compared with the theorical ones [10]

This measurement, although still preliminary and mainly used as a test of validity for the usage of THM in (α, p) reactions, is the first one of the cross-section for the reaction $^{19}F(\alpha,\ p)^{22}Ne$ in the astrophysical energetic region of interest. The same methodology has been used in other recently studied reactions [11] [12] [13] [14] [15].
in the near future, R-Matrix calculations will be used to extract the resonance strength useful for the reaction rate calculations. Such result will be accomplished thanks to the recently developed modified R-Matrix formulation. This is essential to study astrophysical implications of this measurement.

References

[1] Palmerini S, Sergi M L, La Cognata M, Lamia L, Pizzone R G, Spitaleri C, 2013, *ApJ* **764**, 2, 128
[2] La Cognata M, Mukhamedzhanov A M, Spitaleri C, Indelicato I, Aliotta M, Burjan V, Cherubini S, Coc A, Gulino M, Hons Z, Kiss G G, Kroha V, Lamia L, Mrzek J, Palmerini S, Pisko , Pizzone R G, Puglia S M R, Rapisarda G G, Romano S, Sergi M L and Tumino A, 2011, *ApJ Lett.* **739**, L54
[3] Ugalde C, Azuma R, Gorres J, Heil M, Sheller K, Stech E, Tan W and Wiescher M, 2005, *Nuc. Phys. A*, 758, 577
[4] Ugalde C, Azuma R, Coutre A, Gorres J, Lee H Y, Stech E, Strandberg E, Tan W and Wiescher M, 2008, *Phys. Rev. C*, 77
[5] Spitaleri C, 1991, in *Problems of Fundamental Modern Physics, II* : Proceedings, Ed. by R. Cherubini,P. Dalpiaz, and B. Minetti (World Sci., 1991), p. 21
[6] Baur G, Typel S, 2004 *Progress of Theoretical Physics Supplement*, 154, 333-340
[7] Spitaleri C, Mukhamedzhanov A M, Blokhintsev L D, La Cognata M, Pizzone R G and Tumino A, 2011, *Physics of Atomic Nuclei*, **74**, 12, 1725-1739
[8] Tribble R E, Bertulani C A, La Cognata M, Mukhamedzhanov A M and Spitaleri C, 2014, *Reports on Progress in Physics*, **77**, 10
[9] Pizzone R G, Spitaleri C, Mukhamedzhanov A M, Blokhintzev L D, Bertulani C A, Irgaziev B F, La Cognata M, Lamia L and Romano S, 2009, *Phys. Rev. C*, **80**
[10] Bao Z Y, Kappeler F, 1987, *Atomic Data and Nuclear Data Tables*,**36**,411
[11] Tumino A, Spitaleri C, Mukhamedzhanov A M, Typel S, Aliotta M, Burjan V, Gimenez del Santo M, Kiss G G, Kroha V, Hons Z, La Cognata M, Lamia L, Mrazek J, Pizzone R G, Piskor S, Rapisarda G G, Romano S, Sergi M L, Spartá R, 2011 *Phys. Lett. B*, **700**, 2, 111-115
[12] Lamia L, Spitaleri C, La Cognata M, Palmerini S, Pizzone R G, 2012, *A & A*, **541**, A158
[13] Pizzone R G, Spartá R, Bertulani C A, Spitaleri C, La Cognata M, Lalmansingh J, Lamia L, Mukhamedzhanov A M and Tumino A, 2014, *ApJ*, **786**, 112
[14] Tumino A, Spartá R, Spitaleri C, Mukhamedzhanov A M, Typel S, Pizzone R G, Tognelli E, Degl'Innocenti S, Burjan V, Kroha V, Hons Z, La Cognata M, Lamia L, Mrazek J, Piskor S, Prada Moroni P G, Rapisarda G G, Romano S and Sergi M L, 2014, *ApJ*, **785**, 2014
[15] Cherubini S, Gulino M, Spitaleri C, Rapisarda G G, La Cognata M, Lamia L, Pizzone R G, Romano S, Kubono S, Yamaguchi H, Hayakawa S, Wakabayashi Y, Iwasa N, Kato S, Komatsubara T, Teranishi T, Coc A, de Séréville N, Hammache F, Kiss G, Bishop S, Binh D N, 2015, *Phys. Rev. C*, **92**, 1, 015805

8th European Summer School on Experimental Nuclear Astrophysics (Santa Tecla School) IOP Publishing
Journal of Physics: Conference Series **703** (2016) 012017 doi:10.1088/1742-6596/703/1/012017

Low-Energy resonances in the ^{22}Ne(p,γ)^{23}Na reaction directly observed at LUNA

Rosanna Depalo for the LUNA collaboration

Dipartimento di Fisica e Astronomia and INFN Padova, via Marzolo 8, 35131 Padova (Italy)

E-mail: `rdepalo@pd.infn.it`

Abstract. The neon-sodium cycle of hydrogen burning influences the synthesis of the elements between ^{20}Ne and ^{27}Al in AGB stars and classical novae explosions.
The ^{22}Ne(p,γ)^{23}Na reaction rate is very uncertain because of a large number of unobserved resonances lying in the Gamow window.
A new direct study of the ^{22}Ne(p,γ)^{23}Na reaction has been performed at the Laboratory for Underground Nuclear Astrophysics (LUNA) using a windowless gas target and two HPGe detectors. Several resonances have been observed for the first time in a direct experiment.

1. Introduction

The neon-sodium cycle of hydrogen burning contributes to the synthesis of the isotopes between ^{20}Ne and ^{23}Na in Red Giant Branch (RGB) stars, Asymptotic Giant Branch (AGB) stars and classical novae explosions.
The synthesis of sodium in RGB stars is still puzzling. Observations of galactic globular clusters show that the surface abundance of sodium in RGB stars anticorrelates with the oxygen abundance [1]. A possible explanation for this anticorrelation involves the pollution of the interstellar medium with material processed through hydrogen burning reactions at high temperatures in AGB stars. In the hydrogen burning shell of AGB stars, oxygen is efficiently destroyed by the CNO cycle and sodium is mainly produced by the ^{22}Ne(p,γ)^{23}Na reaction of the NeNa cycle [2].
Another scenario where the ^{22}Ne(p,γ)^{23}Na reaction is active are classical novae explosions. A sensitivity study showed that the ^{22}Ne(p,γ)^{23}Na reaction rate uncertainty strongly affects the final abundances of neon, sodium and magnesium isotopes, calling for new experimental efforts on the ^{22}Ne(p,γ)^{23}Na cross section [3].
The ^{22}Ne(p,γ)^{23}Na Gamow window for AGB stars and classical novae extends from 50 to 600 keV. In this energy range, the proton capture on ^{22}Ne is dominated by a large number of resonances. None of the resonances below 436 keV has ever been observed in either direct or indirect experiments (tab. 1). Moreover, the mere existence of the three resonances at 71, 105 and 215 keV is still debated since it has been tentatively reported in [4] but has not been observed in subsequent experiments [5, 6].
As a consequence, the ^{22}Ne(p,γ)^{23}Na reaction rates reported in the widely adopted compilations by NACRE [7] and C. Iliadis et al. [8] are up to three orders of magnitude discrepant in the energy range of interest for hydrogen burning in AGB stars and novae.

Table 1. Summary of literature resonance strengths for ^{22}Ne(p,γ)^{23}Na resonances below 400 keV proton energy.

$E_{res,LAB}$ [keV]	J. Görres (direct) [9]	$\omega\gamma$ [eV] NACRE [7]	Iliadis et al. [8]
29	-	-	$\leq 2.6 \cdot 10^{-25}$
37	-	$(6.8 \pm 1.0) \cdot 10^{-15}$	$(3.1 \pm 1.2) \cdot 10^{-15}$
71	$\leq 3.2 \cdot 10^{-6}$	$\leq 4.2 \cdot 10^{-9}$	-
105	$\leq 6.0 \cdot 10^{-7}$	$\leq 6.0 \cdot 10^{-7}$	-
158	$\leq 1.0 \cdot 10^{-6}$	$(6.5 \pm 1.9) \cdot 10^{-7}$	$(9.2 \pm 3.7) \cdot 10^{-9}$
186	$\leq 2.6 \cdot 10^{-6}$	$\leq 2.6 \cdot 10^{-6}$	$\leq 2.6 \cdot 10^{-6}$
215	$\leq 1.4 \cdot 10^{-6}$	$\leq 1.4 \cdot 10^{-6}$	-
259	$\leq 2.6 \cdot 10^{-6}$	$\leq 2.6 \cdot 10^{-6}$	$\leq 1.3 \cdot 10^{-7}$
291	$\leq 2.2 \cdot 10^{-6}$	$\leq 2.2 \cdot 10^{-6}$	$\leq 2.2 \cdot 10^{-6}$
323	$\leq 2.2 \cdot 10^{-6}$	$\leq 2.2 \cdot 10^{-6}$	$\leq 2.2 \cdot 10^{-6}$
334	$\leq 3.0 \cdot 10^{-6}$	$\leq 3.0 \cdot 10^{-6}$	$\leq 3.0 \cdot 10^{-6}$
369	-	-	$\leq 6.0 \cdot 10^{-4}$
394	-	-	$\leq 6.0 \cdot 10^{-4}$

2. Experimental setup

The Laboratory for Underground Nuclear Astrophysics (LUNA) is located at Gran Sasso National Laboratories, Italy, where the 1400 meters of rocks dominating the laboratory guarantee a reduction of six orders of magnitude in the cosmic muon flux and a reduction of three orders of magnitude in the neutron flux.

The 400 kV electrostatic accelerator provides high intensity (~200 μA) proton or alpha beam. The beam can be delivered either to a solid or to a gas target. Different gamma-ray or particle detectors can be used, depending on the nuclear reaction to be studied [10].

As a first step, a feasibility test was performed using the setup of the previous ^{2}H(α,γ)^{6}Li experiment [11].

For this test, neon gas with natural isotopic composition was used (90.48% ^{20}Ne, 0.27% ^{21}Ne and 9.25% ^{22}Ne). The aim of the test was to study the possible sources of beam induced background, and to have some hints on the sensitivity to the ^{22}Ne+p resonant cross section.

During the test, runs on the ^{22}Ne(p,γ)^{23}Na resonances at 105, 158, 186 and 259 keV have been performed. Despite the use of non enriched gas (only 9.25% ^{22}Ne) and a setup which was not optimized for this measurement, gamma-rays from the 186 keV resonance have been observed in a 12 hours run. This resonance was never observed in previous experiments, and the literature upper limit for the resonance strength is $\omega\gamma < 2.6 \cdot 10^{-6}$ eV. Thanks to the LUNA observation it was possible to provide a lower limit $\omega\gamma \geq 0.12 \cdot 10^{-6}$ eV for the 186 keV resonance strength and thus to improve the literature information on this resonance [12].

Following the feasibility test, the characterization of the setup for the first experimental campaign on the ^{22}Ne(p,γ)^{23}Na resonances was started. In this phase the gas density profile along the beam path was first measured without beam using natural neon gas and a dedicated target chamber. Then the beam heating effect in natural neon gas has been measured for the first time with the resonance scan technique, using the intense and well known ^{21}Ne(p,γ)^{22}Na resonance at 271.6 keV beam energy [13] and a collimated NaI detector [12].

For the ^{22}Ne(p,γ)^{23}Na cross section measurement, a proton beam was delivered to the windowless gas target filled with 99.9% enriched ^{22}Ne.

The gamma-rays emitted in the de-excitation of ^{23}Na were detected by two HPGe detectors collimated at 55° and 90° with respect to the beam direction.
In order to reduce the environmental background at gamma ray energies below 3 MeV, the two detectors were surrounded by a 4 cm thick copper shielding and a 25 cm thick lead shielding.

3. Results

During about five months of data taking, all the resonances between 70 and 334 keV have been investigated. The resonances at 158, 186 and 259 keV have been observed for the first time in a direct experiment. For these resonances, the complete excitation function was measured and then a long run at the energy of maximum yield was performed. New gamma decay modes have also been observed for the three resonances detected.
For the non-detected resonances new upper limits have been measured. The new upper limits are two to three orders of magnitude lower than the previous direct measurement, proving the improvement in sensitivity that can be achieved in underground experiments.
Fig. 1 shows the updated reaction rate for the ^{22}Ne(p,γ)^{23}Na reaction, calculated including the new LUNA results. Literature reaction rates adopted in references [7] and [8] are also shown for comparison.

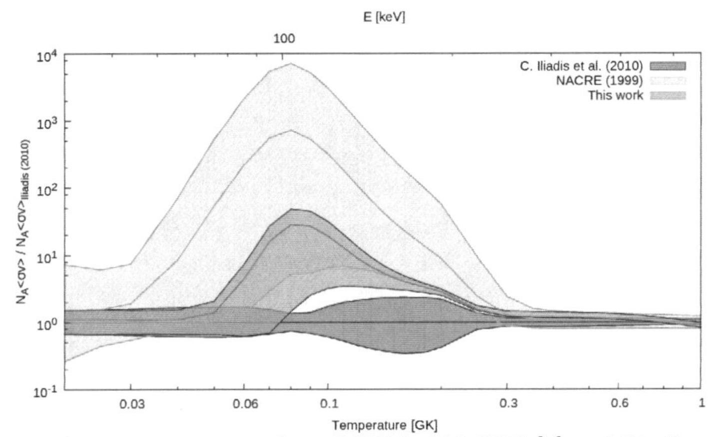

Figure 1. Thermonuclear reaction rate from LUNA, NACRE [7] and Iliadis et al. [8] normalized to [8]. Colored bands represent the reaction rate uncertainties.

The final results will be presented and discussed in detail in a forthcoming publication [15].

References

[1] Carretta E et al. 2009 *Astron. Astrophys.* **505**, 117
[2] Izzard R G et al. 2007 *Astron. Astrophys.* **466** 641
[3] Iliadis C et al. 2002 *Astrophys. J. Suppl.* **142**, 105
[4] Powers J R et al. 1971 *Phys. Rev.* C **4** 6
[5] Hale S E et al. 2001 *Phys. Rev.* C **65** 015801
[6] Jenkins D G et al. 2013 *Phys. Rev.* C **87** 064301
[7] Angulo C et al. 1999 *Nucl. Phys.* A **656** 3
[8] Iliadis C et al. 2010 *Nucl. Phys.* A **841**, 1
[9] Görres J et al. 1982 *Nucl. Phys.* A **385** 57
[10] Broggini C et al. 2010 *Annu. Rev. Nucl. Part. Sci.* **60**, 53
[11] Anders M et al. 2013 *Eur. Phys. J.* A **49** 28
[12] Cavanna F et al. 2014 *Eur. Phys. J.* A **50** 179
[13] Becker H W et al. 1992 *Z. Phys.* A **343** 361
[14] Formicola A et al. 2004 *Phys. Lett.* B **591** 61.
[15] Cavanna F et al. 2015 *arXiv*:1511.05329v1

Lithium and age of pre-main sequence stars: the case of Parenago 1802

M Giarrusso[1 2 3], E Tognelli[4 5], G Catanzaro[3], S Degl'Innocenti[5 6], M Dell'Omodarme[6], L Lamia[1], F Leone[1 3], R G Pizzone[2], P G Prada Moroni[5 6], S Romano[1 2] and C Spitaleri[1 2]

[1]Dip. Fisica e Astronomia, Univ. di Catania, Via S. Sofia 64, I-95123 Catania, Italy
[2]INFN - Laboratori Nazionali del Sud, Via S. Sofia 62, I-95123 Catania, Italy
[3]INAF - Osservatorio Astrofisico di Catania, Via S. Sofia 78, I-95123 Catania, Italy
[4]Dip. Fisica, Univ. di Roma Tor Vergata, Via della Ricerca Scientifica 1, I-00133 Roma, Italy
[5]INFN - Sezione di Pisa, Largo B. Pontecorvo 3, I-57127 Pisa, Italy
[6]Dip. Fisica, Univ. di Pisa, Largo B. Pontecorvo 3, I-57127 Pisa, Italy

E-mail: marina.giarrusso@oact.inaf.it

Abstract. With the aim to test the present capability of the stellar surface lithium abundance in providing an estimation for the age of PMS stars, we analyze the case of the detached, double-lined, eclipsing binary system PAR 1802. For this system, the lithium age has been compared with the theoretical one, as estimated by applying a Bayesian analysis method on a large grid of stellar evolutionary models. The models have been computed for several values of chemical composition and mixing length, by means of the code FRANEC updated with the Trojan Horse reaction rates involving lithium burning.

1. Introduction

Among the observational parameters, surface lithium abundance is definitely of interest for Pre-Main Sequence (PMS) late-type stars, which have deep convective envelopes. As a consequence of the PMS contraction, their interior heats up and when the temperature at the base of the convective zone, T_{BCZ}, reachs that of Li burning, $T_{LB} \sim 2.5\times10^6$ K, lithium is here destroyed via (p,α) reactions. Convection is then responsible for the observed lithium depletion, by bringing the processed material to the stellar surface and the not yet processed one to the inner regions. The efficiency of this process strongly depends on stellar metallicity (mass fractional abundance of elements heavier than helium) and mass. Li-burning is faster in metal-rich stars, since the higher opacity and the consequent deeper convective envelope results in a higher T_{BCZ}. Moreover, fully convective stars reaching T_{LB} (mass 0.06-0.40 $M_\odot$ for solar metallicity, Tognelli et al. 2015) completely destroy the initial lithium content from few tens to hundreds Myr in dependence on the mass: the smaller the mass, the greater the time to reach T_{LB}, the longer the Li-burning timescale. In the interior of more massive stars (0.40-1.20 $M_\odot$) a radiative core develops and grows up as faster as larger is the mass, so that the convective envelope becomes progressively less extended until $T_{BCZ} < T_{LB}$ and the Li-burning is stopped. Since at fixed mass and chemical composition the depth of the convective envelope is age-dependent, in principle the measured surface lithium abundance of PMS late-type stars can be used to derive the stellar age (*lithium age*), if mass and metallicity are known. However, despite the capability of the current stellar models in reproducing the main evolutionary parameters, the still present disagreement between observed and predicted surface lithium content doesn't confirm the validity of this method. Anyway, for PMS stars a better agreement between theory and observations can be obtained, in some cases, just tuning the external convection efficiency (the mixing length parameter), which often results to be lower than the Main Sequence value (Tognelli et al. 2012, hereafter T12).

In this work we have tested the power of the surface lithium abundance to provide an estimation for the age of the PMS, detached, double-lined, eclipsing binary system Parenago 1802 (hereafter PAR 1802), for which surface iron abundance and dynamical masses are available (Table 1).

2. Determination of theoretical and lithium age for PAR 1802

In principle, if we exactly know 1) the metallicity and the mass of a star, 2) a set of observational quantities (e.g. effective temperature, surface gravity, radius) and 3) the observed photospheric lithium abundance, we can a) compute the model with the known values of metallicity and mass, b) within this model, search for the age when the observed parameters (point 2) appear simultaneously (*theoretical age*) and c) independently, search in the model for the age corresponding to the observed lithium abundance (*lithium age*). However, observed parameters (including metallicity and mass) are affected by errors. Therefore, in order to take into account all the uncertainties at the same time, we have applied a Bayesian analysis (Jörgensen and Lindegren 2005) for deriving theoretical estimation of mass (M) and age (τ), by following Gennaro et al. 2012 (hereafter GPT12). The Bayesian Method is a statistical powerful tool which allows to obtain an estimation of stellar parameters (in our case mass and age) by comparing observational evidences of other quantities (in our case effective temperature, surface gravity, radius) with models' predictions for the same quantities, and by taking into account the *a priori* information about metallicity and parameters to be determined, if any. From the resulting model, we have derived the lithium age and compared it with the theoretical one.

We have computed a database of PMS stellar models by means of the Frascati Raphson Newton Evolutionary Code (FRANEC, Degl'Innocenti et al. 2008). Input physics is as in GPT12, limited to the primordial helium abundance Y_p=0.2485 and the helium-to-metal enrichment ratio $\Delta Y/\Delta Z$=2, with the exception of the solar-scaled heavy-element mixture, that is from Asplund et al. (2009). For the initial abundances of ^{6}Li, Be, B we have adopted the values given in Prantzos (2012), while we have set the logarithmic initial ^{7}Li abundance A(^{7}Li)[1]=3.2 (e.g. Jeffries 2000). It has to be noted that the accuracy of theoretical evolutionary models in predicting surface abundance of light elements largely relies on both the adopted mixing length parameter and the nuclear reaction rates. The former can be arbitrarily set in the code as input, e.g. on the basis of considerations related to the evolutionary stage of the star in exam (T12). Regarding to the charged-particle-induced reactions at astrophysical energies (~keV), the cross sections are underestimated by direct measurements because of the Coulomb barrier (~MeV) between the interacting ions, and overestimated by low-energy extrapolation procedures on direct measurements made at higher energies because of the electronic screening phenomena due to the electron clouds surrounding the interacting nuclei. The Trojan Horse Method (THM; Baur 1986; Spitaleri et al. 2011) is a powerful indirect technique which allows to measure the bare-nucleus factor at astrophysical energies without invoking both Coulomb penetrability and electron screening effects (see e.g. Tumino et al. 2007). We have adopted nuclear reaction rates from NACRE compilation and JINA REACLIB database, but for the following ones derived with the THM: d(d,p)t and d(d,n)^{3}He (Tumino et al. 2014), ^{6}Li(p,α)^{3}He (Lamia et al. 2013), ^{7}Li(p,α)^{4}He (Lamia et al. 2012). We have obtained models for different metallicities Z: 0.00700,

[1] A(^{7}Li) = 12 + $log N_{Li}/N_H$ where N_{Li} and N_H indicate respectively the lithium and hydrogen numerical abundance.

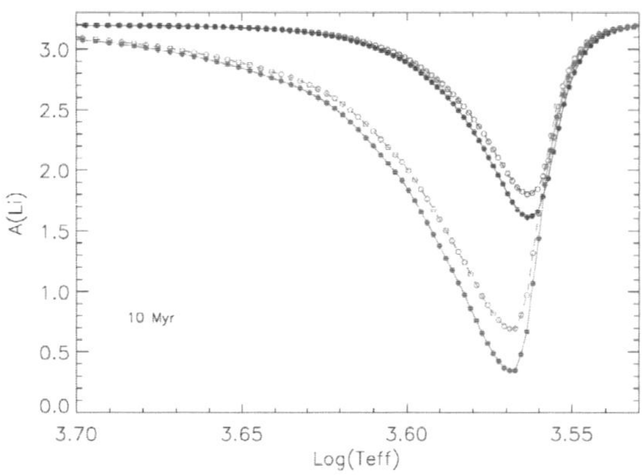

Figure 1. Logarithmic surface ^{7}Li abundance is plotted against T_{eff}, at the same age (10 Myr), for two classes of models with solar metallicity and mixing-length parameter α=1.00 (purple), α=1.74 (red). Each class has been computed with different inputs for cross sections of reaction involving lithium, i.e. empty circles for THM and filled circles for NACRE (Angulo et al.1999).

Table 1. Input stellar parameters for PAR 1802 A: M_{dyn}=0.391±0.032 $M_\odot$, $T_e[K]$= 3675±150, log$g$=3.55±0.04, R=1.73±0.02 $R_\odot$, A(^{7}Li)=1.48±0.21. For B component, in the same order: 0.385±0.032, 3365±150, 3.61±0.04, 1.62±0.02, 0.57±0.22 (Cargile et al. 2008, Gómez Maqueo Chew et al. 2012). [Fe/H]=-0.01±0.04 (D'Orazi et al. 2009). A(^{7}Li) are according to Pavlenko and Magazzú (1996) NLTE computations. Stellar (τ_{star}), system (τ_{sys}) and lithium (τ_{Li}) ages are in Myr.

PAR	α = 1.00				α = 1.74			
1802	M[$M_\odot$]	τ_{star}	τ_{sys}	τ_{Li}	M[$M_\odot$]	τ_{star}	τ_{sys}	τ_{Li}
A	$0.40^{+0.02}_{-0.04}$	$1.05^{+0.11}_{-0.11}$	$1.10^{+0.09}_{-0.08}$	$12.6^{+0.05}_{-0.05}$	$0.39^{+0.04}_{-0.02}$	$0.75^{+0.08}_{-0.07}$	$0.79^{+0.07}_{-0.05}$	$13.0^{+0.05}_{-0.05}$
B	$0.39^{+0.02}_{-0.04}$	$1.16^{+0.13}_{-0.14}$		$12.6^{+0.05}_{-0.05}$	$0.38^{+0.02}_{-0.02}$	$0.85^{+0.08}_{-0.08}$		$13.0^{+0.05}_{-0.05}$

0.00800, 0.00900, 0.01000, 0.01291 (solar value), 0.01550 and 0.01800. For each Z-value, we have computed models with mixing-length α=1.00 (suitable for PMS stars, T12) and α=1.74 (the solar-calibrated value). In order to obtain a high precision in the Bayesian analysis, we have computed a very fine grid of models in the mass range [0.20, 2.50] $M_\odot$ as it follows. The models have initially been obtained with a spacing of 0.25 $M_\odot$. Then we have interpolated the grid with a spacing of 0.0125 $M_\odot$ and 0.005 Myr in the full mass range.
An interesting comparison for surface lithium predicted by models obtained with different inputs is shown in Figure 2.
Physical parameters used in the analysis are in Table 1 together with the Bayesian results.

3. Results and Conclusions

Figure 2 shows Bayesian results for α=1.00 and 1.74. In both cases (Table 1) the theoretical age (~1 Myr) indicates a system too young to burn lithium, while the lithium age (~13 Myr) indicates a system with an efficient lithium burning process. As a conclusion from our analysis, differences between theoretical ages and lithium ages, at least in the case of PAR 1802, cannot be ascribed to the mixing length parameter.

References

Angulo C et al. 1999 *Nucl. Phys.* A **656** 3
Asplund M, Grevesse N, Sauval A J and Scott P 2009 *Ann. Rev. Astron. Astrophys.* **47** 481
Baur G 1986 *Phys. Lett.* B **178** 135

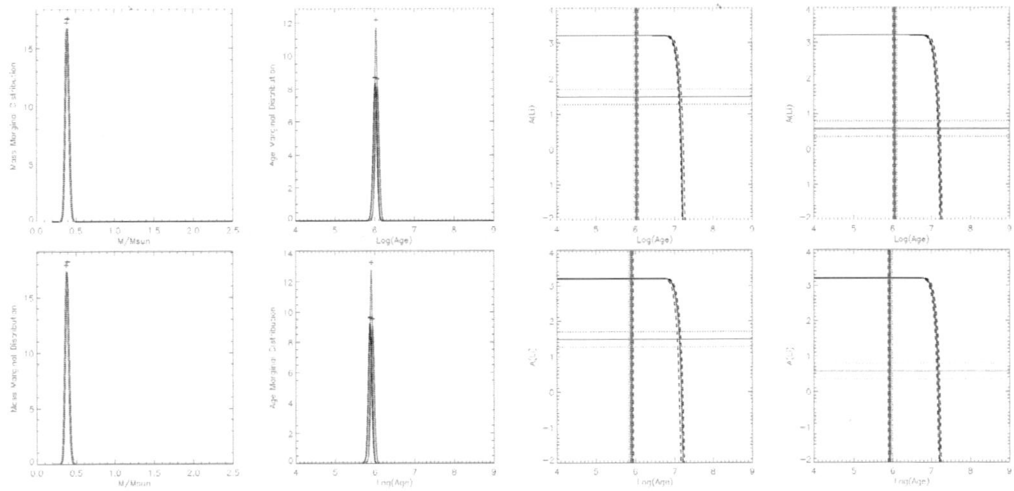

Figure 2. Bayesian results for α=1.00 (top panels) and α=1.74 (bottom panels). From left: mass marginal distribution of primary (blue) and secondary (red) component, age marginal distribution of primary (black), secondary (blue) and system (red), lithium depletion pattern of primary and secondary component in black solid line, dashed the propagation of mass error from confidence interval. Blue vertical solid line marks the bayesian age of the single star, dashed the associated error from the confidence interval. Red vertical solid line marks the system age (coevality), dotted the associated error from the confidence interval. Red horizontal solid line represents the observed lithium abundance, dotted the associated error. Horizontal line in marginal distribution represents the confidence interval.

Cargile P A, Stassun K G and Mathieu R D 2008 *Astrophys. Journ.* **674** 329
Degl'Innocenti S, Prada Moroni P G, Marconi M and Ruoppo A 2008 *Astroph. Sp. Sci.* **316** 25
D'Orazi V, Randich S, Flaccomio E, Palla F, Sacco G G and Pallavicini R 2009 *Astron. Astrophys.* **501** 973
Gennaro M, Prada Moroni P G and Tognelli E 2012 *Mont. Not. Roy. Astron. Soc.* **420** 986
Gómez Maqueo Chew Y, Stassun K G, Prsa A, Stempels E, Hebb L, Barnes R, Heller R and Mathieu R D 2012 *Astrophys. Journ.* **745** 58
Jeffries R D 2000 *Proc. from Astronomical Society of the Pacific Conf.* (25-28 May 1999 Mondello, Palermo) vol 198, ed R Pallavicini et al (ASP) p 509
Jørgensen B R and Lindegren L 2005 *Astron. Astrophys.* **436** 127
Lamia L, Spitaleri C, La Cognata M, Palmerini S and Pizzone R G 2012 *Astron. Astrophys.* **541** 158
Lamia L, Spitaleri C, Pizzone R G, Tognelli E, Tumino A, Degl'Innocenti S and Prada Moroni P G 2013 *Astrophys. Journ.* **768** 65
Pavlenko Ya V and Magazzú A 1996 *Astron. Astrophys.* **311** 961
Prantzos N 2012 *Astron. Astrophys.* **542** 67
Spitaleri C, Mukhamedzhanov A M, Blokhintsev L D, Cognata M L, Pizzone R G and Tumino A 2011 *Phys. Atom. Nucl.* **74** 1725
Tognelli E, Degl'Innocenti S and Prada Moroni P G 2012 *Astron. Astrophys.* **548** 41
Tognelli E, Degl'Innocenti S and Prada Moroni P G 2015 *Mont. Not. Roy. Astron. Soc.* **449** 3741
Tumino A et al 2007 *Phys. Rev. Lett.* **98** 252502
Tumino A et al 2014 *Astrophys. Journ.* **785** 96

Journal of Physics: Conference Series **703** (2016) 012019 doi:10.1088/1742-6596/703/1/012019

Oxygen isotopic ratios in RGB & AGB stars

Ghina M. Halabi[1,2] & Mounib El Eid[2]

[1]Institute of Astronomy, University of Cambridge, Madingley Road, Cambridge CB3 0HA, UK

[2]American University of Beirut, Bliss Street 11-0236, Beirut 1107 2020, Lebanon

E-mail: gmh@ast.cam.ac.uk

Abstract. Stars in the mass range 1.2-7$M_\odot$, with solar-like initial composition are evolved from the zero age main sequence till the early AGB phase. Their predicted oxygen isotopic ratios are compared to observationally inferred values in red giants to investigate how well standard evolutionary calculation reproduces available observations and to illustrate the role of convective mixing. We find that extra mixing beyond the convective boundary determined by the Schwarzschild criterion is needed to explain the observational oxygen isotopic ratios, particularly in low mass stars. The effect of recent determinations of proton capture reactions and their uncertainties on $^{16}O/^{17}O$ is also shown.

1. Introduction

The evolution to the red giant branch (RGB) follows the ignition of shell H-burning surrounding the He core. The released energy flux causes the envelope to expand and a deep convective envelope develops, leading to a change in the surface composition of the star. In this contribution, an observed sample of red giants [1] is used to estimate the masses by matching their observationally derived effective temperatures and bolometric magnitudes to the values obtained from the evolutionary tracks in the HR diagram. This is possible since the stars are not pulsating Mira variables (see [2] for details). We then compare their calculated oxygen abundances to those inferred from observations. This comparison is useful in order to improve the treatment of convective mixing in the stellar interiors. In particular, the abundance profile of ^{17}O exhibits a steep gradient within the central region of the star, at the position of maximum convective penetration. This renders the ^{17}O surface abundance sensitive to the depth of the convective envelope, the stellar mass, as well as to the nuclear reaction rates involved in the ON cycle. Our results suggest the need for extra mixing below the edge of the convective envelope as determined by the Schwarzschild criterion in order to achieve a better agreement with observations. The effect of several determinations of p-capture ON reaction rates on $^{16}O/^{17}O$ is investigated.

This work is organized as follows. Section 2 describes the calculations. Some results are presented in Section 3. Concluding remarks are given in Section 4.

2. Model Calculations

The evolutionary sequences presented in this work are obtained using the stellar evolution code HYADES as described in [3]. Details on the calculation and results can be found in [2]. Mass-loss is included using semi-empirical rates adjusted to the global parameters of the star. Overshooting, or mixing beyond the Schwarzschild boundary is introduced according to the hydrodynamic simulations by [4], where an exponentially decaying diffusion coefficient is introduced at the convective boundary.

3. Results

3.1. Evaluation of the stellar masses

Using the bolometric magnitudes and effective temperatures of the star sample observed by [1], evolutionary tracks of (1.2-7)$M_\odot$ stars are used to evaluate the stellar masses as shown in Fig. 1.

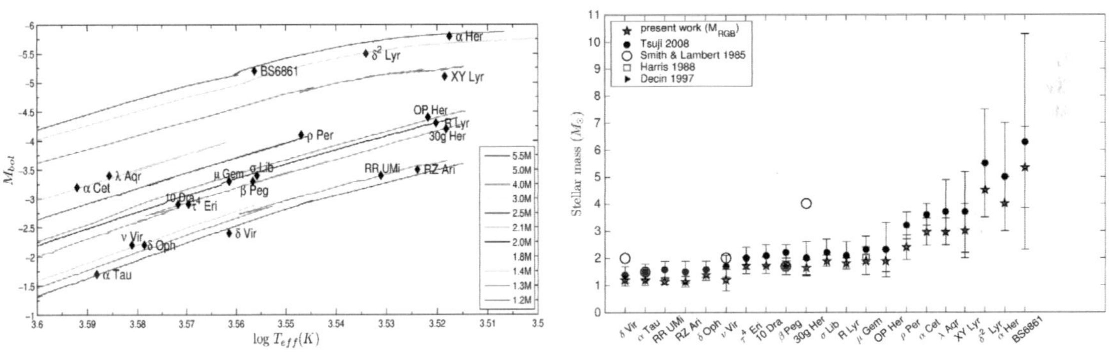

Figure 1. Left panel: Evolutionary tracks shown with the observed sample of red giants. Right panel: The stellar masses reported with the error bars in addition to other evaluations as shown, in solar units.

3.2. Oxygen Surface Abundances

We use the $^{16}O/^{17}O$ of the observed red giants, whose masses are determined, to constrain the extra mixing required on the RGB. The left panel of Fig. 2 shows the predicted $^{16}O/^{17}O$ ratios in the standard case (no extra mixing), along with other predictions and available observations. In the case of low-mass stars, the predicted $^{16}O/^{17}O$ ratio is higher than most data points obtained from observations. The results of standard calculation by other groups shown in the figure indicate a similar behavior. This shows that more ^{17}O needs to be mixed to the surface, which may be achieved by extra mixing or overshooting at the bottom of the convective envelope. Applying overshooting described in Sect. 2, our calculations show that $f = 0.125$ provides the best estimate for the overshooting efficiency in low mass stars. The right panel of Fig. 2 shows the results with overshooting. The low observational $^{16}O/^{17}O$ values in low mass stars can now be better fitted within the error bars on the stellar masses.

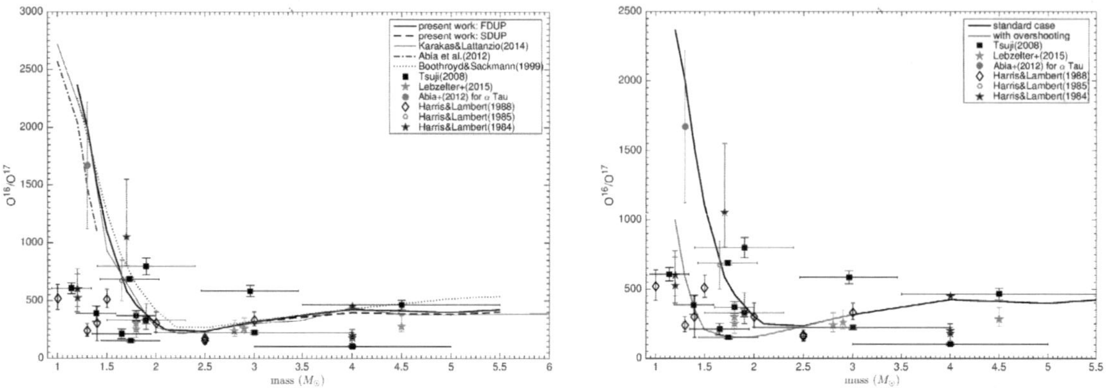

Figure 2. Left panel: $^{16}O/^{17}O$ ratios in the standard case. Calculations by other groups and observations are shown for comparison. Right panel: The ratios obtained with overshooting at the bottom of the envelope with $f = 0.125$. The solar value is 2620.

3.3. Effect of the ON p-capture reaction rates

To investigate the effect of the ^{17}O production and destruction rates on the $^{16}O/^{17}O$ ratios, four different evaluations of the proton-capture reactions $^{16}O(p,\gamma)^{17}F$, $^{17}O(p,\gamma)^{18}F$ and $^{17}O(p,\alpha)^{14}N$ are used. In particular, we use the compilations by [5] (SE14), [6] (SA13), [7] (IL10) and [8] (CH07). The ratios obtained with the four sets in the standard case are shown in the left panel of Fig. 3, exhibiting very similar values in the considered mass range. None provides a reasonable agreement between model predictions and observations unless overshooting is included.

The effect of rate uncertainties is also worth exploring. Since the SA13 compilation is based on a Monte Carlo simulation and the rates have statistically well-defined uncertainties, the ratios are calculated using the recommended, high and low rates, where the rate boundaries correspond to a 95% coverage probability. The $^{16}O/^{17}O$ ratios show a large sensitivity to these uncertainties, as shown in the right panel of Fig. 3. It is found that the discrepancy between predictions and observations becomes less pronounced when the uncertainty on the involved reaction rates is considered.

4. Conclusions

A sample of observed red giants was considered. Their masses were obtained using extended evolutionary tracks and their predicted oxygen abundances were compared to available observations. Overshooting based on the mixing of chemical elements by diffusion is needed to the reconcile theoretical and observational $^{16}O/^{17}O$ ratios, particularly in low mass stars. Moreover, we find that the experimentally suggested uncertainty on relevant reaction rates may reduce the existing discrepancy but does not exclude the need to invoke overshooting.

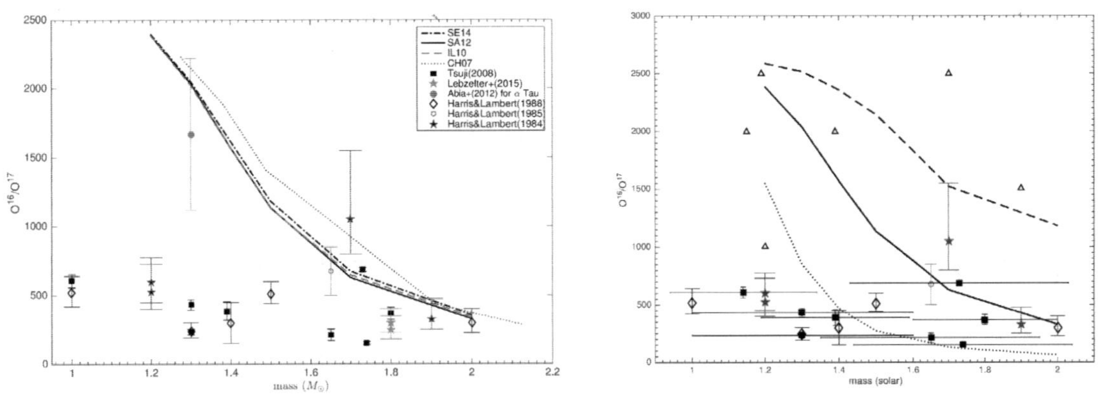

Figure 3. Left panel: $^{16}O/^{17}O$ ratios in the standard case using four different sets of reaction rates (see text for details). Right panel: The SA13 set of rates: recommended (solid line), high rates (dotted) and low rates (dashed).

4.1. Acknowledgments

GMH thanks Dr. C. Spitaleri and the conference organizers for the generous support.

References

[1] Tsuji, T. 2008, A&A 489, 1271
[2] Halabi, G. M., El Eid, M., 2015, MNRAS, 451, 2957
[3] El Eid, M. F., The, L.-S., & Meyer, B. 2009, Space Sci. Rev., 147, 1
[4] Freytag, R., Ludwig, H., & Steffen, M. 1996, A&A, 313, 497
[5] Sergi, L. et al. 2014, AIPC, 1594, 201S
[6] Sallaska, A. L., Iliadis, C., Champange, A. E., Goriely, S., Starrfield, S., Timmes, F. X. 2013, ApJS 207, 18
[7] Iliadis, C., Longland, R., Champagne, A., Coc, A. 2010, Nucl.Phys. A841, 251-322
[8] Chafa, A. et al., 2007, PRC 75, 035810
[8] Smith, V. V. & Lambert D. L., 1985, ApJ, 294, 326
[9] Harris, M. J., Lambert, D. L. & Smith, V.V. 1988, ApJ, 325, 768
[10] Decin, L., Cohen, M., Eriksson, K., Gustafsson, B., Huygen, E., Morris, P., Plez, B., Sauval, J., Vandenbussche, B., Waelkens, C. 1997, ESASP, 419, 185D
[11] Boothroyd, A. I. & Sackmann, I.-J. 1999, ApJ, 510, 232
[12] Abia, C., Palmerini, S., Busso, M., & Cristallo, S. 2012, A&A, 548, A55
[13] Karakas, A. & Lattanzio J. 2014, PASA, 31, e030
[14] Lebzelter, T., Straniero, O., Hinkle, K.H., Nowotny, W. & Aringer, B. 2015, A&A, 578A, 33L
[15] Harris, M. J. & Lambert, D. L., 1984a, Ap,J 281, 739
[17] Harris, M. J., Lambert, D. L. & Smith, V. V. 1985, ApJ, 299, 375

The study of neutron-rich nuclei production in the region of the closed shell N=126 in the multi-nucleon transfer reaction ^{136}Xe+^{208}Pb

K.Novikov[1], I.M.Harca[1,10,11], E.M.Kozulin[1], S.Dmitriev[1], J.Itkis[1], G.Knyazheva[1], T.Loktev[1], L.Corradi[2], J.Valiente-Dobon[2], E.Fioretto[2], D.Montanari[2], A.M.Stefanini[2], E.Vardaci[3], D.Quero[3] G.Montagnoli[4], F.Scarlassara[4], E.Strano[4], G.Pollarolo[5], J.Piot[5], T.Mijatoviĉ[6], S.Szilner[6], D.Ackermann[7], G.Chubarian[8] and W.H.Trzaska[9]

[1]Flerov Laboratory of Nuclear Reactions, Joint Institute for Nuclear Research, Dubna, Russia
[2]Instituto Nazionali di Fisica Nucleare, Laboratori Nazionali di Legnaro, Legnaro, Italy
[3]Dipartamento di Scienze Fisiche and INFN (INFN-Na), Napoli, Italy
[4]Dipartimento di Fisica, Universit di Padova, and Instituto Nazionale di Fisica Nucleare, Padova, Italy
[5]GANIL, CEA/DSM-CNRS/IN2P3, Bd Henri Becquerel, BP 55027, F-14076 Caen Cedex 5, France
[6]Ruer Boşkoviĉ Institute, and University of Zagreb, Croatia
[7]GSI Helmholtzzentrum fr Schwerionenforschung, Darmstad, Germany
[8]Cyclotron Institute, Texas A & M University, College Station, TX 77843-3366, USA
[9]Accelerator Laboratory of University of Jyväskylä (JYFL), Jyväskylä, Finland
[10]Horia Hulubei National Institute for R & D in Physics and Nuclear Engineering (IFIN-HH), Bucharest - Magurele, Romania
[11]Facultatea de Fizica, Universitatea din Bucuresti, Bucharest, Romania

E-mail: `harca@jinr.ru`

Abstract. The unexplored area of heavy neutron rich nuclei is extremely important for nuclear astrophysics investigations and, in particular, for the understanding of the r-process of astrophysical nucleogenesis. For the production of heavy neutron rich nuclei located along the neutron closed shell N=126 (probably the last "waiting point" in the r-process of nucleosynthesis) the low-energy multi-nucleon transfer reaction ^{136}Xe+^{208}Pb at E_{lab}=870MeV was explored. Due to the stabilizing effect of the closed neutron shells in both nuclei, N=82 and N=126, and the rather favorable proton transfer from lead to xenon, the light fragments formed in this process are well bound and the Q-value of the reaction is nearly zero.
Measurements were performed with the PRISMA spectrometer in coincidence with an additional time-of-flight (ToF) arm on the +20 beam line of the PIAVE-ALPI accelerator in Legnaro, Italy. The PRISMA spectrometer allows identification of the A, Z and velocity of the projectile-like fragments (PLF), while the second arm gives access to the target-like fragments (TLF). Details on the experimental setup and preliminary results are reported.

Journal of Physics: Conference Series **703** (2016) 012021 doi:10.1088/1742-6596/703/1/012021

1. Introduction

The transfer of many nucleons was suggested as a feasible route to synthesize heavy nuclei on the neutron-rich side of the stability line [1],[2],[3], giving rise to a renewed interest in the study of transfer processes.

Within this work, the low-energy collision of ^{136}Xe with ^{208}Pb was explored. The stabilizing effects of the neutron closed shell N = 82 of ^{136}Xe and N = 126 of ^{208}Pb, together with the specific trend of the Q-values of all possible mass transfer channels [4], make the chosen system a very good candidate. Protons may experience a higher mobility due to of the neutron closed shell in the projectile and target, and the transfer of several protons from Xe to Pb may lead to very neutron-rich nuclides [3].

2. Experimental Configuration and Data Analysis

The experiment was carried out at the Laboratori Nazionali di Legnaro (LNL) by bombarding a ^{136}Xe beam at E_{lab}= 870 MeV, provided by the Tandem-ALPI complex, onto a 200 μg/cm^2 thick (and 99.9% enriched) ^{208}Pb target, backed on 20 μg/cm^2 Carbon foil. The target was placed at 90° with respect to the beam direction.

PRISMA is a large-acceptance magnetic spectrometer, operating at LNL, designed for identification of reaction products of heavy ion collisions consisting of a Micro Channel Plate detector (MCP), a focusing quadrupole, a dipole magnet, a focal Multiwire Parallel Plate detector (MW-PPAC) and, finally, the ionization chambers (IC) (for ΔE-E measurement). The PRISMA set-up allows identification of the A and Z of the projectile-like fragments (PLF) of reaction within a resolution of 1/220 and 1/60, respectively.

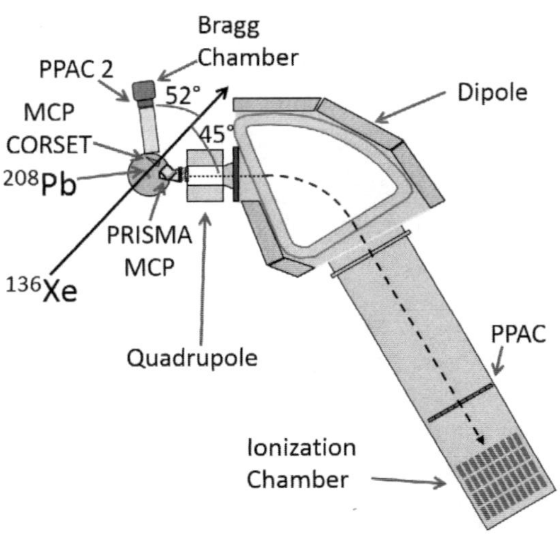

Figure 1. Schematic arrangement of the set-up.

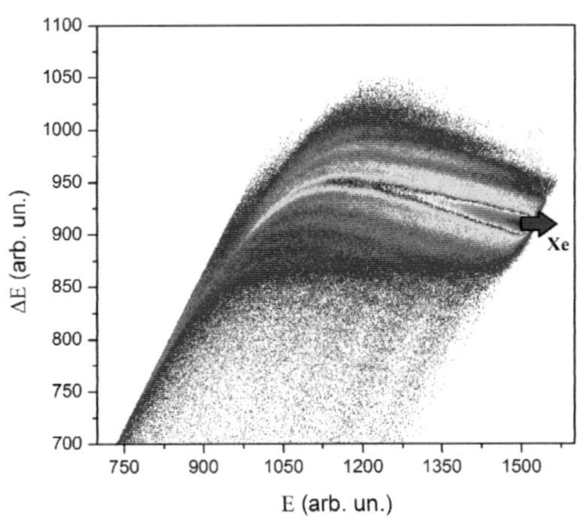

Figure 2. E vs ΔE matrix for the reaction ^{136}Xe + ^{208}Pb, as measured by the ionization chamber of PRISMA. The most intense track corresponds to the Xe beam, while the region of interest corresponds to Z(Xe)+6.

For our present case study, the PRISMA setup had in its implementation a second arm, composed of a CORSET [5] compact Micro Channel Plate (MCP) start detector followed by a position-sensitive stop detector and, finally, a Bragg ionization chamber. A schematic view of the setup is shown in figure 1.

The experimental setup provides several methods of separating between good events with different properties. Firstly, the identification of the nuclear charge, Z, is performed within the energy and energy losses measured in the ionization chamber. Figure 2 displays the ΔE vs E matrix for the ^{136}Xe+^{208}Pb system where the Z separation is clearly visible. Secondly, once the ions' trajectories are uniquely reconstructed on an event-by-event basis, for each selected track in the previously mentioned matrix, one can disentangle different charge states. Because the magnetic fields are known, one may calculate the radius of curvature, R, of the ions inside the dipole and their velocities v. Figure 4 shows the $R{\cdot}v$ vs E matrix for the enclosed band of $Z = 54$.

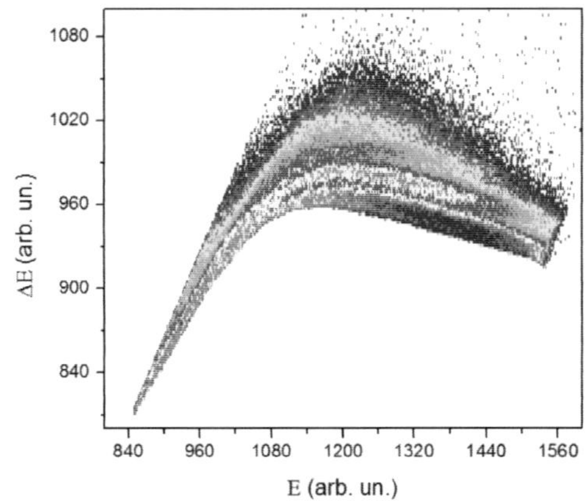

Figure 3. Same matrix as in Figure 2, displaying only events corresponding to Z>Z(Xe).

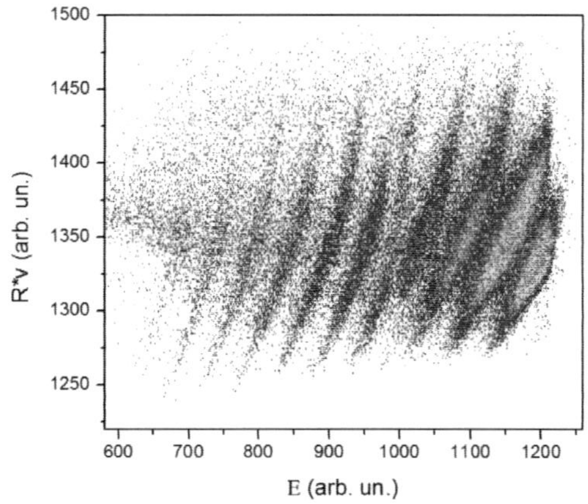

Figure 4. E vs $R \cdot v$ matrix for the Xe ions. The most intense band corresponds to the most probable charge state $q = 40^+$.

3. Conclusions

We report the current status on the production of heavy neutron-rich nuclei located along the neutron closed shell N=126 illustrated in the low-energy multi-nucleon transfer reaction ^{136}Xe+^{208}Pb, performed at E_{lab}=870MeV.

The probability of light particle emission for this type of reaction is negligible, thus the cross section for production of the neutron-rich heavy nuclei, such as Pt, is within experimental reach by using multiple coincident gating. In order to estimate experimentally the cross section for production of new neutron rich heavy nuclei, the mass distribution (with a $\approx \pm 1 - 2$ u resolution) will be obtained by using the time of flight between the PRISMA MCP and PPAC and the fully reconstructed trajectory of ions inside the spectrometer. The analysis is in progress.

References

[1] V.V.Volkov, Phys. Rep., 44, 93 (1978).
[2] C.H.Dasso et al.,Phys. Rev. Lett., 73 1907 (1994).
[3] V. Zagrebaev and W. Greiner, Phys. Rev. Lett. 101, 122701 (2008).
[4] E.M. Kozulin et al., Phys. Rev. C86(2012) 044611.
[5] E.M. Kozulin et al., ISSN 0020-4412, Instruments and Experimental Techniques 51 (2008) 44.

Effects of Cell Windows on TwinSol Beams

Kimmy Cushman
SUNY College at Oneonta, New York, United States
E-mail: cushkk80@suny.oneonta.edu

Abstract. In order to study reactions with unstable nuclei, radioactive-ion beams must be used. One method for producing radioactive beams is the TwinSol experimental setup at the University of Notre Dame. At TwinSol, stable and unstable isotope beams bombard a gas target, where one atmosphere of gas must be confined from the surrounding vacuum. Thin foil windows are used to contain the gas in the cell. In order to optimize the quality of secondary beams from TwinSol, it is necessary to understand and minimize the effects of energy loss and straggling in the windows. This work is the beginning of a process to improve the TwinSol design so that secondary beams produced with heavier ions such as Oxygen, Fluorine, and Neon can be pursued.

1. Introduction

One method to probe nuclear structure is the bombardment of light nuclei with a heavy, radioactive nuclei. However, in the lab, we must replicate the reaction using inverse kinematics because radioactive isotopes cannot be used as a target. In order to isolate the gas target from the vacuum, gas cell windows must be used. This investigation will focus on the effect of the cell windows on TwinSol experiments.

One of the most important properties of gas cell window is that is essentially invisible to the beam. To approach this optimization, a very thin foil may be used. However, as the strength of a foil is proportional to its thickness, very thin foils will not withstand the vacuum. In order to produce a practical amount of reaction products, one atmosphere of gas is typically contained within the cell, and such a pressure gradient leads to stress on the windows. Therefore, a window material has to be relatively strong.

Another key feature of the optimal window material is that it has a low stopping power, dE/dx. Since the stopping power is proportional to Z^2, a low Z material will absorb less energy from the beam. Similarly, a low density material will yield lower energy straggling. When the beam passes through the window, it loses energy due to the electric and nuclear forces experienced between the beam and the material. We would like to diminish effects of stopping power and reduce energy straggling in the beam in order to increase the resolution of the beam and thus reduce the error of the calculated cross sections of reactions.

Furthermore, window materials which are thermally conductive will be able to dissipate the heat of the beam, which is focused to a diameter of a few millimeters. Having a low thermal conductivity, however, can be supplemented by having a high melting or degradation temperature. It is also important to note that the initial pressure of the gas cell may greatly increase due to heating of the gas by the beam, yielding greater stress on the window.

Journal of Physics: Conference Series 703 (2016) 012022 doi:10.1088/1742-6596/703/1/012022

Material	Thickness μm	Tensile Strength MPa	Density mg/cm^3	Thermal Cond. W/mk	Melting/Degredation Temperature °C
Titanium	5	246-370	4.50	21	1668
Molybdenum	2.5	550-650	10.8	138	2623
Aramid [3]	4 and 12	392-490	1.5	n/a	300-350
Aluminized Mylar	24	Mylar 196 Aluminum 40-50	1.39 2.70	¡1 237	254 660

Table 1:
Materials used in vacuum and beam experiment.

2. Experimental Design

The gas window used in previous TwinSol experiments were 5 μm Titanium foils. Possible window material alternatives to Titanium are evaluated based on the following properties: tensile strength, mailability, thickness, density, thermal conductivity, melting/degradation temperature, availability, cost. In order to understand durability of materials subject to a ion beam, we used experimental methods as well as theoretical simulations and calculations. Stopping powers of each beam and window material combination were calculated using the SRIM program [1], and these were used to calculate the power and energy implanted in each of the windows. See figure 1. Also, for each material, the maximum current load was calculated using the VTL (Virtual Target Lab) program [2].

The central experiment of this investigation was the vacuum and beam test, which were meant to mimic typical TwinSol conditions. See table 1 for materials tested with the given properties.

In the experiment, a 5+ charge state Carbon beam was accelerated to 42.5 MeV by the FN-tandem accelerator at the University of Notre Dame. ^{4}He gas filled the cells to approximately atmospheric pressure. The pressure inside the gas cells is to be monitored by a digital pressure gauge, and recorded throughout the experiment. The procedure for each test is to begin at a low current of 50 nA for tuning, and then increase the current until the window breaks or significantly leaks.

3. Results

The results of the experiment were very useful in providing the first metrics against which to compare theoretical results. Given the many parameters which must be controlled and the time constraints of the experiment, the materials must be compared with caution.

Window Material	Integrated Current pnA · hr
5 μm Titanium	1.27
2.5 μm Molybdenum†	723
4 μm Aramid	35.0
12 μm Aramid	273
24 μm Mylar no osc.	16.7
24 μm Mylar w osc.	16.4

Table 2:
Titanium window used to align the beam and received the smallest integrated current. The Molybdenum windows received the highest integrated current by a large factor, without failure†. The integrated current of the Aramid and Mylar windows is the total current reached before failure.

3.1. Theoretical Results

Using SRIM stopping power tables, energy loss was calculated for each of the window materials given initial beam conditions. Figure 1 contrasts the fractional energy loss of a range of Carbon beam energies in Titanium and Aramid windows against the thickness of the material.

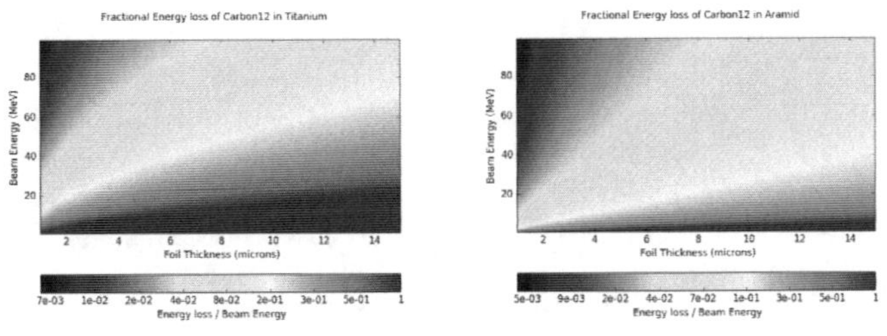

(a) (b)

Figure 1: Calculations using SRIM stopping power tables. Contrast the fraction of beam energy loss in Titanium (a) versus the less dense Aramid (b). Where the gradient is deep red, indicating a fractional loss of 1, all of the beam energy is lost because at low beam energies and high window thicknesses, the beam does not have enough energy to escape the window, so it acts as a beam stop. Even when the thickness of Aramid is double that of Titanium, the energy loss in Aramid is less.

4. Discussion and Future Research

Preliminary results of this investigation indicate that Molybdenum and Aramid may be suitable alternatives to Titanium for TwinSol gas cell windows. If Molybdenum is to be used, a very thin foil, such as the 2.5 μm, should be used because Molybdenum is a high Z element, compared to Titanium (42 vs. 22). Although this increases its stopping power, Molybdenum has a much higher melting point and thermal conductivity, so it may be advantageous in having a long lifetime of use in experiments. An alternative with opposite characteristics was found to be the Aramid foils. Having a low density of 1.5 g/cm^3, this material has a low stopping power, which is optimal for maintaining the resolution of primary and secondary TwinSol beams. However, it may not have a long lifetime of use under high beam currents. Therefore, the specificity of each TwinSol experiment may determine which window material may be ideal.

4.1. Future Work

This investigation was meant to be the beginning of a process to improve the Twinsol design and understand how the gas cell windows affect the primary and secondary beams. From this point, outputs from SRIM and VTL may be used with the accompanying codes for calculations so that more alternatives may be explored. The experimental procedure can be also followed in future tests to compare the material performances. In future experiments of this type, it could be beneficial to implement an infrared heat camera to image the windows and monitor their temperature on line. With this apparatus, theoretical equilibrium pressures from VTL could be compared to experiment.

Although the conditions required for this experiment were considered for TwinSol, understanding the effects of gas cell windows on nuclear beams is not unique to experiments conducted at the University of Notre Dame. For any experiment where a gas target is required, methods for investigating alternative window materials would be beneficial.

References

[1] J. Ziegler, SRIM (The Stopping Ranges of Ions in Matter), software available at `http://www.srim.org`
[2] S. Kasemann, VTL (The Virtual Target Lab), University of Cologne 1996
[3] Toray Industries, Inc. `http://www.toray.com/products/fibers/fib_012.html`

Journal of Physics: Conference Series **703** (2016) 012023 doi:10.1088/1742-6596/703/1/012023

Complex analysis of scattering 1p-shell nuclei in the framework of coupled channel method

M Nassurlla[1,3], N Burtebayev[1], A Duisebayev[1], J Burtebayeva[1], C Spitaleri[2], A Urkinbayev[1], K Rusek[4], E Piasecki[4], S Kliczewski[5], A Trzcinska[4], S B Sakuta[6], I Boztosun[7], S V Artemov[8], L I Galanina[9]

[1] Institute of Nuclear Physics, 050032, Almaty, Kazakhstan
[2] National Institute of Nuclear Physics, Sezione di Catania, I-95125, Catania, Italy
[3] Kazakh National University, 050038, Almaty, Kazakhstan
[4] Heavy Ion Laboratory of Warsaw University, 02-093, Warsaw, Poland
[5] H. Niewodniczański Institute of Nuclear Physics Polish Academy of Sciences, 31-342, Krakow, Poland
[6] Russian Research Center "Kurchatov Institute", 123182, Moscow, Russia
[7] Akdeniz University, Faculty of Science, Department of Physics, 07058, Antalya, Turkey
[8] Institute of Nuclear Physics, 100214, Tashkent, Uzbekistan
[9] Skobeltsyn Institute of Nuclear Physics of Moscow State University, 119991, Moscow, Russia
E-mail: *morzhic@mail.ru*

Abstract. The scattering process on 1p-shell nuclei, having the cluster structure, can be seen in the anomaly increasing of cross sections for large angles. Most often, this increasing of cross sections is connected with mechanism of transfer of clusters or nucleons. The study of the α-cluster transfer mechanism in the elastic scattering of ^{20}Ne ions on ^{16}O nuclei is important for investigation burning process in evolution of the Universe immediately after the Big-Bang. Therefore new experiment on the heavy ion accelerator (Warsaw University) was carried out with a significant expansion of the range of angles up to 170^0 in center mass system at E_{Lab}=50.0 MeV. Data analysis of angular distribution was performed in framework of the optical model and coupled channel method. The optimal parameters of the optical potential were obtained and the spectroscopic factor was obtained 1 for ^{20}Ne as α +^{16}O.

1. Introduction

Better knowledge about spectroscopic factor could provide us useful information about nuclear structure of nuclei and reaction mechanism. It is well known that, the spectroscopic factor is related to the preformation probability of the cluster configuration in a nucleus. Usually, there are different methods to extract the value of the spectroscopic factors from the experimental measurements. The spectroscopic factors play significant role especially in elastic transfer reactions which result in increasing the differential cross sections at backward angles and consequently coupled channel calculations are required to describe this process. Transfer reactions usually take place at low energies. The growth of the cross sections at backward angles was observed which is typical for α-cluster structured nuclei [1]. This increase could be interpreted to be due to the contribution of mechanism of α-cluster transfer [2]. The interaction of these nuclei has been represented by Stock *et al.* [3]. In the present work reanalysis of differential cross section for ^{20}Ne+^{16}O system was performed at E_{Lab}=50 MeV.

2. Experiment

The measurements were carried out at the extracted beam ^{20}Ne from the K = 160 cyclotron of the Heavy Ion Laboratory, University of Warsaw. The beam energy was 2.5 MeV/n. The charged particles were detected and identified by four ΔE-E counter telescopes installed in the ICARE experimental chamber. The overall energy resolution was about 700 keV. Self-supporting targets of aluminum-oxide with a thickness of 150 μg/cm^2 were mounted in the chamber. In order to determine thicknesses of targets there was used the scattering chamber, installed in the protons ion wire of the accelerator "UKP-2-1", were measured energy losses of the protons beam during the passage through target. For this purpose the reaction $^{27}Al(p,\gamma)^{28}Si$ was used with narrow resonance with E_R=992 keV and with detection of gamma-quanta with E_γ=1779 keV. The thicknesses of the fine films (see figure 1) were determined by analyzing the shift of the resonance E_R=992 keV in the reaction $^{27}Al(p,\gamma)^{28}Si$ which takes place at the expense of protons energy losses in the thickness of target, using table values of brake quantities "S(E_p) (MeV*cm^2/g), there were determined thicknesses of fine films (see figure 1). Such method allows to determine thicknesses of films in the interval of (10÷100) μg/cm^2 with the accuracy not worse than 5%.

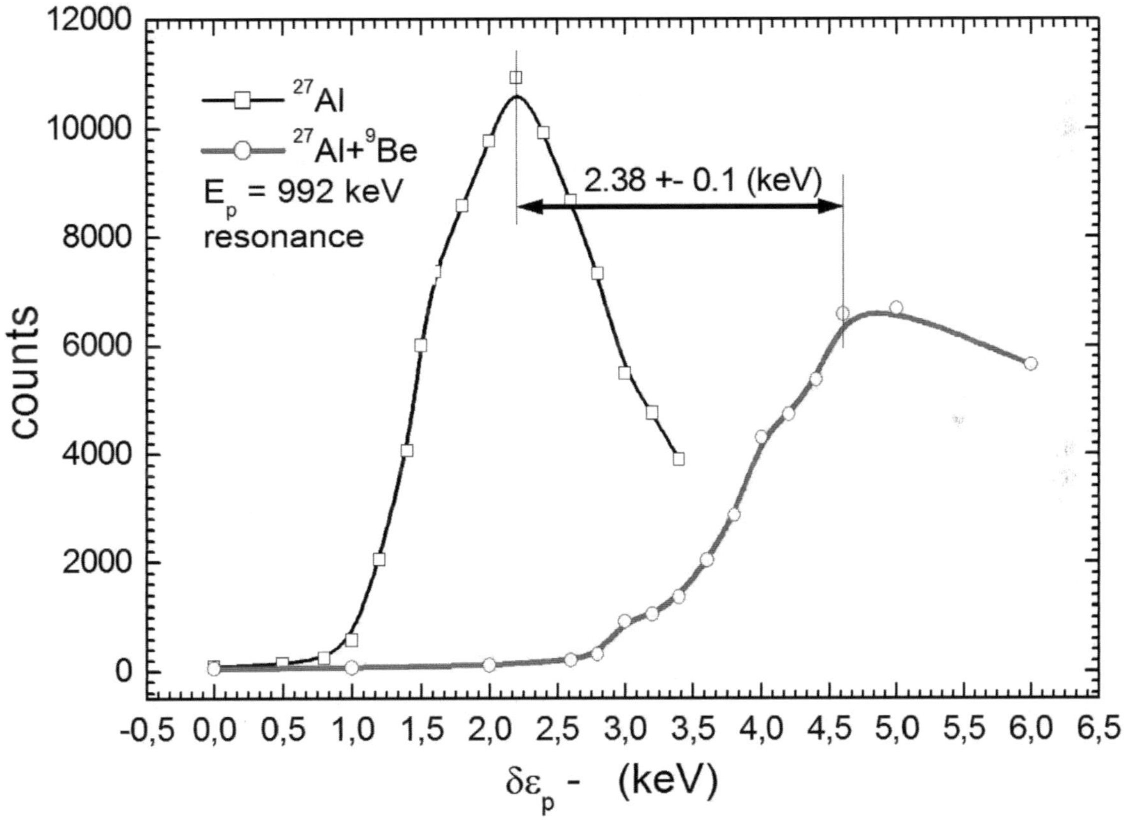

Figure 1. The shift of the 992 keV resonance in reaction ^{27}Al (p, γ) ^{28}Si due to the loss of energy of protons in the film thickness of ^9Be. The thickness of the target for ^9Be is 10.75 g / cm^2 with an accuracy of 5%.

The ΔE-E spectrum of the ^{20}Ne (^{16}O, ^{16}O) ^{20}Ne reaction products was obtained via ROOT (see figure 2). Our experimental measurements showed scattering pattern at forward hemisphere while, a significant increase in differential cross section at backward hemisphere is observed as shown in figure 3.

Journal of Physics: Conference Series **703** (2016) 012023 doi:10.1088/1742-6596/703/1/012023

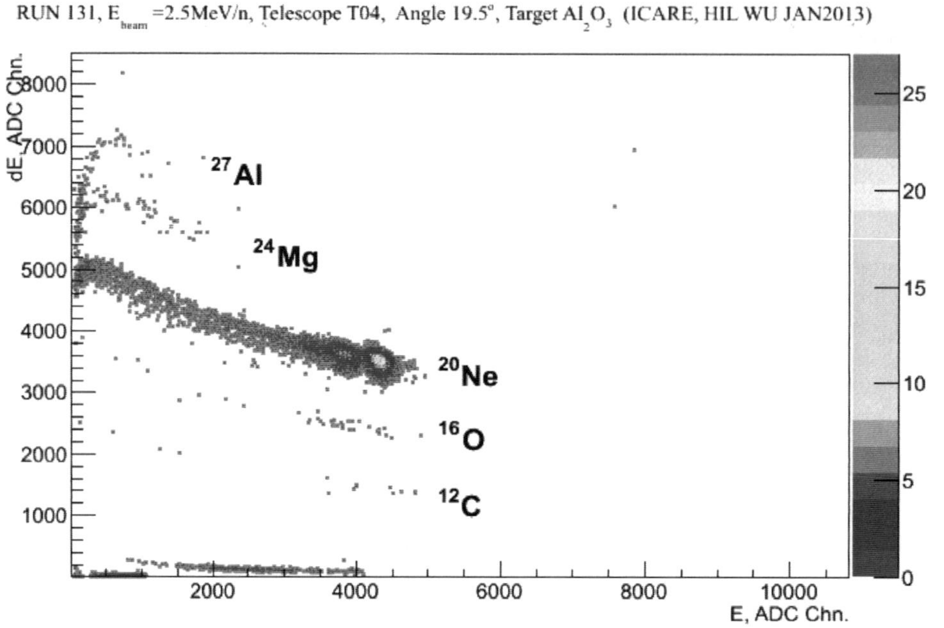

Figure 2. ΔE-E spectrum of the ^{20}Ne (^{16}O, ^{16}O) ^{20}Ne reaction products.

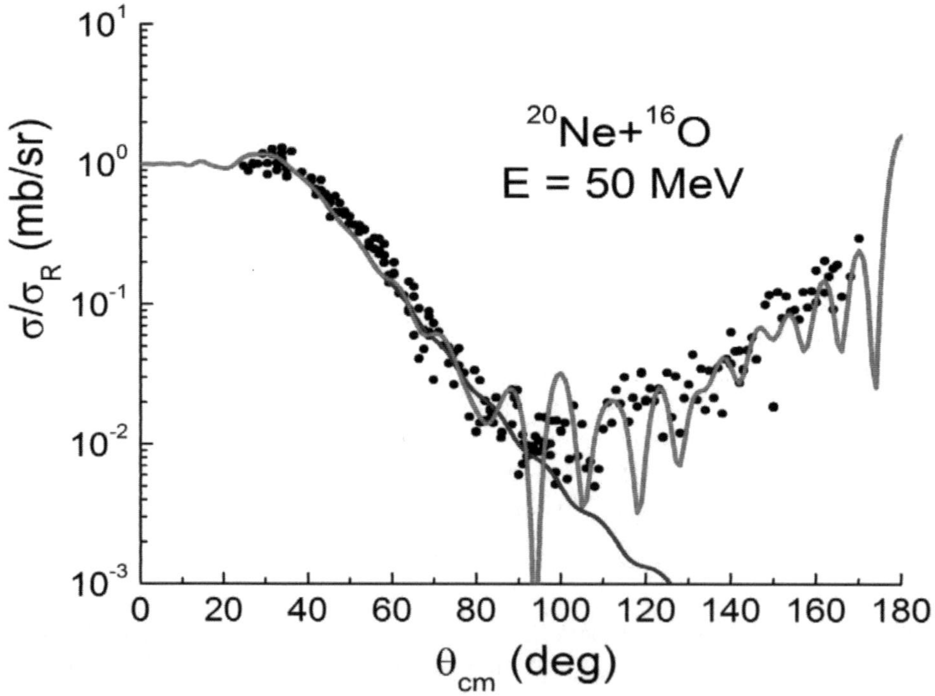

Figure 3. Angular distribution of the elastic scattering of ^{20}Ne on ^{16}O at the 50 MeV energy. Circles are experimental data; the blue line is the optical model prediction, and the red line represents the coupled reaction channels calculation by code FRESCO with taking into account the exchange mechanism of the α-cluster transfer.

3. Data analysis

Experimental data of elastic scattering at forward hemisphere were analyzed within the framework of optical model (OM). For all calculations on OM, the Woods-Saxon form factor was used for both the real and imaginary potential

$$U = V + iW \tag{1}$$

$$V = V_o\left[1 + exp\frac{(r - R_r)}{a_r}\right]^{-1} \tag{2}$$

$$W = W_o\left[1 + exp\ \frac{(r - R_i)}{a_i}\right]^{-1}, \tag{3}$$

V_o and W_o, a_r and a_i, R_r and R_i being the depth, diffuseness and radii of the real and imaginary potentials, respectively. The radii are expressed in terms of the mass numbers A_1 and A_2 of the nuclei involved given by

$$R = r_o\left(A_1^{1/3} + A_2^{1/3}\right). \tag{4}$$

Parameters of optical potential (OP) were selected to achieve the best agreement between theoretical and the experimental angular distributions (see table). The description of experimental data is shown in figure 3 (represented by the blue line). It can be seen that OM fails to describe the differential cross section of elastic scattering in backward hemisphere.

Table. Potential Parameters.

E (MeV)	V_0 (MeV)	R_r (fm)	a_r (fm)	W_0 (MeV)	R_i (fm)	a_i (fm)	SF
50.0	100.0	1.20	0.49	35.0	1.31	0.49	1

Proposed nuclear system has been analyzed earlier by Stock *et al.* [3] and it shows an oscillatory structure at the intermediate angles and a dramatic rise of the cross section at backward hemisphere. The optical model description could not give a reasonable explanation of cross section behavior at intermediate and large angles. In a recent work, Yang and Li [4] have studied the same nuclear system in the energy range of $E_{c.m:}$ = 24.5 to 35.5 MeV in framework of folding model which assumes ^{16}O nucleus consist of four-α-particle and $\alpha+^{16}O$ the configuration for the ^{20}Ne. They have obtained a sufficient description of the experimental data for these energies. Our first attempt to describe the $^{20}Ne+^{16}O$ system in framework of optical and coupled channel models was done in [5]. The key of that work we used a mixed combination of the potentials (deep real potential with a sum of Woods-Saxon typed surface and volume imaginary potentials). Such calculation also was made in other work for different nuclear system [6]. In that paper, the significant rise of the scattering cross section of α particles on ^{9}Be nuclei can be reproduced with using a set of optical potential with deep real part that not taking into account the possible contribution of elastic and inelastic transfer of α-cluster. Therefore, the relevant description of data in backward hemisphere needs to be taken into account the α-transfer mechanism for the system

$$A + (A' + x) \rightarrow A' + (A + x) \tag{5}$$

at $A = A'$ allows reproducing the cross sections rise at large angles (see figure 4).

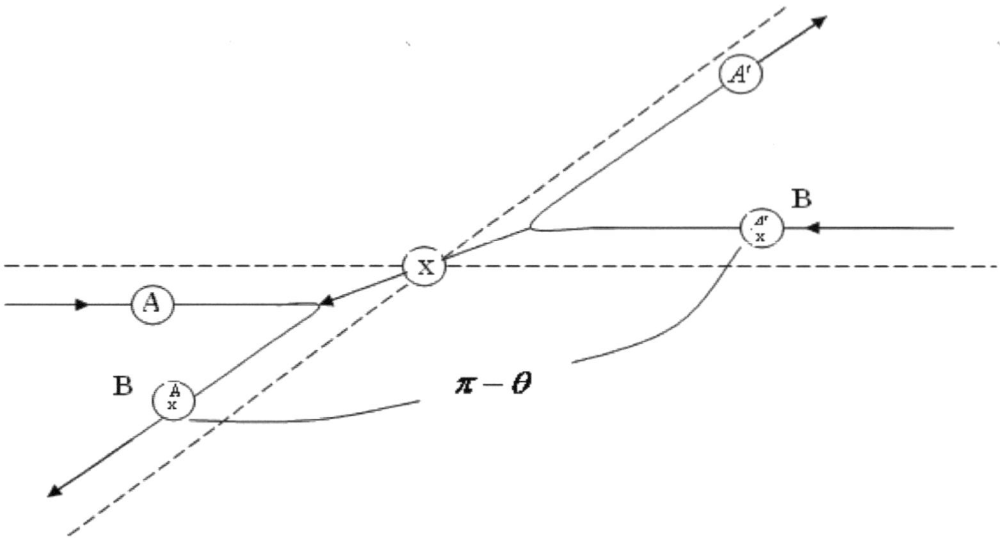

Figure 4. Schematic representation for the transfer reaction.

The contribution of the α-transfer mechanism was calculated within of the framework of the coupled channel method via code FRESCO [7] using the optical model potential parameters from the table. Result of the calculation is shown in the figure 3 by the red line. The spectroscopic factor $S = 1$ for ^{20}Ne → ^{16}O + α was extracted from agreement of experimental data with the coupled channel calculations. The better fitting between the experimental data and coupled channel method calculations was given for the oscillations at the intermediate angles comparing to Stock *et al.* [3]. In addition, our experimental data were obtained in a wide angular range, up to 170^0.

4. Summary

This work was devoted to the interaction of α-cluster structure type nuclei such as ^{20}Ne+^{16}O system at E_{Lab}=50.0 MeV. The experimental elastic ^{20}Ne+^{16}O scattering data have been obtained with a significant expansion of the range of angles up to 170^0 in c.m. Were obtained the optimal parameters of the optical potential. The α-cluster transfer mechanism calculated in framework of the coupled channel method can only give a reasonable description in whole range of angles including the intermediate region which responses for the oscillation structure. The extracted spectroscopic factor is 1 for ^{20}Ne as α +^{16}O.

[1] Burtebayev N *et al.* 2011 *Nucl. Phys.* **A859** 29
[2] Burtebayev N *et al.* 2013 *The VII European Summer School on Experimental Nuclear Astrophysics (Catania)* p 43
[3] Stock R *et al.* 1976 *Phys. Rev.* **C14** 1824
[4] Yang Yong-Xu and Li Qing-Run 2011 *Phys. Rev.* **C84** 014602
[5] Burtebayev N *et al.* 2015 *J. Phys.:Conf. Ser.* **590** 012056
[6] Peterson R J 1982 *Nucl. Phys.* **A377** 41
[7] Thompson I J 1988 *Comput. Phys. Rep.* **7** 167

Journal of Physics: Conference Series **703** (2016) 012024 doi:10.1088/1742-6596/703/1/012024

Towards the study of ^{2}H(p, γ)^{3}He reaction in the Big Bang Nucleosynthesis energy range in LUNA

Izabela Kochanek
Laboratori Nazionali del Gran Sasso, INFN
E-mail: iza.kochanek@lngs.infn.it

Abstract. The Big Bang Nucleosynthesis began a few minutes after the Big Bang, when the Universe was sufficiently cold to allow deuterium nuclei to survive photo-disintegration. The total amount of deuterium produced in the Universe during the first minutes depends on the cosmological parameters (like the energy density in baryons, $\Omega_b h^2$, and the effective neutrino number, N_{eff}) and on the nuclear cross sections of the relevant reactions. The main source of uncertainty in the deuterium estimation comes from the ^{2}H(p, γ)^{3}He cross section.
Measurements of Cosmic Microwave Background (CMB) anisotropies obtained by the Planck satellite are in very good agreement with the theoretical predictions of the minimal ΛCDM cosmological model, significantly reducing the uncertainty on its parameters. The Planck data allows to indirectly deduce with very high precision the abundances of primodial nuclides, such as the primodial deuterium fraction ^{2}H/H = $(2.65 \pm 0.07) \cdot 10^{-5}$ (68% C.L.).
The astrophysical observations in damped Lyman-α systems at high redshifts provide a second high accuracy measurement of the primodial abundance of deuterium ^{2}H/H = $(2.53 \pm 0.04) \cdot 10^{-5}$ (68% C.L.).
The present experimental status on the astrophysical S-factor of the ^{2}H(p, γ)^{3}He reaction in the BBN energy range, gives a systematic uncertainties of 9%. Also the difference between *ab-initio* calculations and experimental values of S_{12} is at the level of 10%.
In order to clarify the actual scenario, a measurement of ^{2}H(p, γ)^{3}He cross section with a precision of a few percent in the 70-400 keV energy range is planned at LUNA in 2016. A feasibility test of the measurement has been performed in October 2014, giving the preliminary results on the cross section. The experimental setup for the test and final measurement campaign will be presented.

1. Introduction

The Big Bang Model is supported by three observational evidences: the cosmic expansion, the Cosmic Microwave Background (CMB) radiation and Big Bang Nucleosynthesis (BBN). BBN predicts the primordial abundances of the light cosmological elements, like ^{4}He, ^{2}H, ^{3}He and ^{7}Li that are produced through twelve key nuclear reactions.

The abundance of light isotopes depends on the energy density in baryons $\Omega_b h^2$ and on the number of neutrno families, N_{eff}. $\Omega_b h^2$ is measured from the angular power spectrum of the CMB temperature anisotropies. A precise value of this parameter was updated by the recent mesurements of Planck satellite to $\Omega_b h^2 = 0.02207 \pm 0.00033$ [1]. Assuming ΛCDM, the Planck constraints on the baryonic density can be translated into a prediction for the primordial deuterium fraction $^2H/H = (2.65 \pm 0.07) \cdot 10^{-5}$ (68% C.L.) [2]

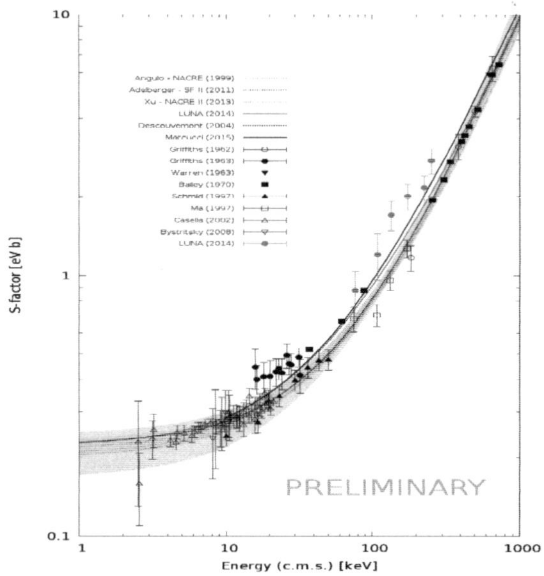

Figure 1. The astrophysical S(E) experimental data together with theoretical *ab initio* calculations and preliminary results from the test run.

The abundance of ^{2}H is directly measured on known deuterium absorption-line systems, including some new data from very metal-poor Lyman-α systems at high redshifts. The ^{2}H/H ratio in these clouds is thought to reflect the primordial abundance and the adopted average value is $^2H/H = (2.53 \pm 0.04) \cdot 10^{-5}$ (68% C.L.) [3].

An independent estimation on the cross section of the radiative capture reaction ^{2}H(p, γ)^{3}He can be extracted from a combined analysis of Planck data and of the deuterium abundance in Lyman-α systems, assuming the standard cosmological model. The cross section for proton-deuterium capture is an important parameter not only in models of BBN, but also in stellar hydrogen burning and deuterium depletion in low-mass protostars, which is believed to be essential for their genesis and evolution.

2. Existing experimental data

The main uncertainty for standard BBN calculations of deuterium comes from the rate of radiative capture reaction ^{2}H(p, γ)^{3}He. This is the first reaction in a chain that rapidly burn ^{2}H into ^{3}He and eventually ^{4}He. The experimental data of proton-deuterium radiative capture reaction are measured in a wide energy range from 2.5 keV to 1.75 MeV.

The low energy limit of the ^{2}H(p, γ)^{3}He cross section is well known thanks to the recent results of the underground experiment LUNA [4]. During BBN, the relevant energy range is around E_{cm} = 30 - 300 keV and for such energies, the uncertainty on σ(E) is at the level of 6 - 10 %. There are several data sets for this reaction in the BBN window: however, some of these measurements need detailed consideration. It has been suggested that the Griffiths'63 [5] and Bailey'70 [6] experiments used incorrect stopping powers and, consequently, their low energy behaviour is 15% too high. The Schmid'97 [7] data set suffer from poor energy resolution, with typical uncertainties greater than 10%. Ma'97 [8] data has the systematic uncertainties estimated to be ±9% at 40≤ E_{cm} ≤110 keV. Moreover, the discrepancy between *ab initio* calculations [9], [10] and experimental fit of S(E) from the Solar Fusion II is at the level of 10%.

In order to clarify the scenario, a very precise measurement of $^2H(p,\gamma)^3He$ in the energy range E_{cm} = 70 - 400 keV is planned at LUNA.

3. Experimental setup and results from the test run

In October 2014, a campaign to test the feasibility of the setup for the precise measurement was held with an High Purity Germanium (HPGe) detector. The experimental setup was based on a windowless gas target chamber filled with deuterium at 0.3 mbar (without recirculation) and a HPGe (at 90° with respect to the beam direction). No lead castle and anti-radon box were present, since the natural background is negligible in the RoI of γ's produced by the ^{2}H(p, γ)^{3}He reaction.

The S(E) was measured at different energies: 112.5 keV, 163.9 keV, 199.5 keV, 259.5 keV, 340 keV and 380 keV with an average current of 150 μA. The results can be seen on Figure 1, together with existing data points.

During the test run, no neutron production or neutron/deuterium implantation was noticeable. The fluorine contamination occurs as a narrow peak at high energy for 380 keV run due to $^{19}F(p,\alpha\gamma)^{16}O$ strong resonance.

In order to obtain the S(E) reported here, a Monte Carlo detection efficiency estimation was used. In the future, a procedure for the experimental determination of the detection efficiency will be adopted.

4. Final measurement campaign

The possibility to study the $^2H(p,\gamma)^3He$ in the solar energy range (2.5 - 22 keV) with good accuracy has been already demonstrated at LUNA [4]. The final measurement of $^2H(p,\gamma)^3He$ reaction in the BBN energy range is planned at the beginning of 2016. This measurement will be divided into two phases: Phase I (BGO) and Phase II (HPGe).

In Phase I, a 4π BGO detector will reduce the (systematic) uncertainty of the detector response for the angular distriubution of the emitted γ-rays. The detection efficiency can be determined by performing a dedicated calibrations and by precise Monte Carlo simulations. With the proposed setup, the expected counting rate will be of the order of $10^4 - 10^5$ events/hour in the considered energy range under the normal conditions (I=100 μA, p=0.1 mbar, l=17.7 cm).

Phase II will be devoted to the study of the angular distribution of emitted γ-rays. This will be acomplished by using a HPGe detector facing the gas target in close geometry within a new target chamber. The expected counting rate will be of the order of $10^3 - 10^4$ events/day under the same running conditions of Phase I.

5. Conclusions

The feasibility test showed that a precise measurement of the $^2H(p,\gamma)^3He$ cross section is possible at LUNA. In order to minimise the systematic uncertainties, the campaign will be done with two detectors. The improved accuracy in the ^{2}H(p, γ)^{3}He cross section measurement will be fundamental for the BBN. Moreover, the study of γ angular distribution will provide a useful tool for nuclear theoretical models.

References

[1] P.A.R. Ade et al. for Planck Collaboration, *Astron.Astrophys.* **571** (2014) A16
[2] E. Di Valentino et al., 2014 *Phys. Rev. D* **90** 2
[3] R. Cooke et al., 2014 *Astrophys. J.*, **781**, 1, 31
[4] C. Casella for the LUNA Collaboration, 2002 *Nucl.Phys. A*, **706** 203
[5] G. M. Griffiths et al., 1963, *Canadian Journal of Physics*, **41**, 4, 402-411
[6] G. M. Bailey et al., 1970, *Canadian Journal of Physics*, **48**, 24, 3059-3061
[7] G. J. Schmid et al., 1996 *Phys. Rev. Lett.*, **76** 17, 3088-3091
[8] L. Ma et al., 1997, *Phys. Rev. C*, **55**, 2, 588–596
[9] L. E. Marcucci, 2006, *Nucl. Phys. A*, **777**, 111
[10] L. E. Marcucci, 2015, *arXiv:1510.07877v1*

Journal of Physics: Conference Series **703** (2016) 012025 doi:10.1088/1742-6596/703/1/012025

Study of ^{16}O(^{12}C,α^{2} Ne)α for the investigation of carbon carbon fusion reaction via the Trojan Horse Method.

G.G. Rapisarda[1,2], C. Spitaleri[1,2], C. Bordeanu[3], Z. Hons[4], G.G. Kiss[1], M. La Cognata[1], J. Mrazek[4], C. Nita[3], D. Pantelica[3], H. Petrascu[3], R.G. Pizzone[1], S. Romano[1,2], T. Szücs[5], L. Trache[3,6], A. Tumino[1,7], G. Velisa[3].

[1] *INFN Laboratori Nazionali del Sud, Catania, Italy*
[2] *Dipartimento di Fisica e Astronomia Università di Catania, Italy*
[3] *NIPNE, Bucharest, Romania*
[4] *Nuclear Physics Institute of ASCR, Rez near Prague, Czech Republic*
[5] *ATOMKI, Debrecen, Hungary*
[6] *Cyclotron Institute Texas A M University, College Station, TX, USA*
[7] *Università degli studi di Enna Kore, Facoltà di Ingegreria e Architettura, Enna, Italy*

mail: `grapisarda@lns.infn.it`

Abstract. arbon carbon fusion reaction represents a nuclear process of great interest in astrophysics, since the carbon burning is connected with the third phase of massive stars (M > M$_\odot$) evolution. In spite of several experimental wor s, carbon carbon cross section has been measured at energy still above the Gamow window moreover data at low energy present big uncertainty. In this paper we report the results about the study of the ^{16}O(^{12} ,α^{20}Ne)α reaction as a possible three body process to investigate ^{12} (^{12} ,α)^{20}Ne at astrophysical energy via Trojan orse Method (T M). This study represents the rst step of a program of experiments aimed to measure the ^{12} +^{12} cross section at astrophysical energy using the T M.

1. Introduction

The ^{12} ^{12} reaction is of great interest both in nuclear physics, producing the rst evidence of nuclear molecular phenomena, as well as in astrophysics. In particular in this last eld carbon burning is connected with the third phase of massive star evolution in stars with M > M$_\odot$ 1 . ^{12} ^{12} reaction rate is a fundamental parameter to determine the so called M$_{up}$, that is, the minimum mass of a star for carbon ignition. Stars with M$_{up}$ evolve into hite dwarf, while stars with M > M$_{up}$ conclude their life as core collapse Supernovae 2 . The core carbon burning ta es place in a temperature range of T .5 1. G . For a temperature of .5 G , the corresponding Gamow energy for the ^{12} ^{12} fusion is E_G 1.5 ± .3 Me . Accurate determination of the carbon burning reaction rate re uires very precise measurement of the ^{12} ^{12} cross section down to this energy, well below the oulomb barrier. In spite of the ey role of carbon fusion reactions in understanding stellar evolution, experimental data available (3 and references therein) cover down to carbon carbon center of mass energy $_{cm}$ 2 Me , that is at the higher edge of the Gamow pea . Moreover experimental data below $_{cm}$ 3 Me are rather uncertain 2 . p to now people calculate the reaction rate by means of the extrapolation from data at higher energy, but this procedure can lead to wrong result since the ^{12} ^{12} excitation function is characteri ed by resonant structures also at the low energy of

Journal of Physics: Conference Series **703** (2016) 012025 doi:10.1088/1742-6596/703/1/012025

astrophysical interest 3 . In this framewor new and accurate experimental data, down to the astrophysical energies, are strongly re uired.
A possible way to measure ^{12} ^{12} excitation function down to astrophysical energies, overcoming the problems connected to the direct measurement, mainly the strong suppression of the cross section due to the oulomb barrier, is by using indirect methods. Among of them the Tro an orse Method (T M) 5, 6, 7 is a consolidated method applied successfully for the study of nuclear reactions of astrophysical interest. In this paper, we report on the measurement of the ^{16} (^{12} , α^{20}Ne)α in uasi free inematic condition to study the possibility to apply the T M to this three body reaction for the indirect study of the ^{12} (^{12} , ^{20}Ne) α reaction, that is the alpha channel of the carbon carbon fusion. This represents the rst measurement of an experimental program devoted to the study of the carbon burning via the T M.

2. The Trojan Horse Method

The T M is a powerful techni ue that allows to extract a two body reaction cross section, A $x \rightarrow c$ C , down to the low energies of astrophysical interest by selecting the uasi free brea up channel of a suitable three body reaction A $a \rightarrow c$ C s. Nucleus a is selected to have a high probability for a cluster con guration $x \oplus s$. The A a interaction induces the nucleus a to brea up into x and s. Selecting the uasi free inematic conditions it is assumed that s acts as a spectator while x interacts with the nucleus A leading to the astrophysically relevant two body reaction. The A a reaction is induced at energies higher than the oulomb barrier, in this way the brea up can ta e place inside the nuclear eld and accordingly the A. x interaction ta es place without su ering the oulomb barrier suppression and the electron screening. Moreover, than s to $x \oplus s$ inter cluster motion, the T M allows to measure the two body cross section in a wide energy using a single beam energy.
T M has been widely applied to study nuclear reactions involved in several astrophysical scenario including reactions induced by unstable nuclei and by neutrons , , 1 , 11, 13, 14 .

3. Experimental set-up

In the present case ^{16} (^{12} , α^{20}Ne)α had been selected as possible three body reaction for the indirect study of the ^{12} (^{12} , ^{20}Ne)α reaction via T M. For the rst time ^{16} has been selected as Tro an orse nucleus for its possible ^{12} $\oplus$ α con guration. The experiment too place at the oria ulubei National Institute of hysics and Nuclear ngineering (IFIN) Bucharest, Romania. The M Tandem accelerator provided a 25 Me ^{16} beam with a spot si e on the target of about 2 mm and an intensity of about nA. A natural carbon target, 1 μg cm^2 thic , was used to induce the ^{16} ^{12} reaction. nergy and position of the outgoing particles were detected using six charge partition position sensitive silicon detectors (S) in a symmetric con guration to double the number of collected events. S s 1,2,3 covered the angular ranges 13° 26°, 41° 55° and 65° ° respectively S s 4,5,6 were placed on the other side with respect to the beam axis, covering the same angular ranges. article identi cation was carried out with Δ techni ue. In particular, two ioni ation chambers (I) lled by 1 (average pressure 23 mb) were placed in front of S 1 and S 4. The two telescopes (I S) were devoted to the neon detection and identi cation. The experimental set up described above had been set to measure the ^{12} (^{12} ,α ^{20}Ne) excitation function in a wide range 3 Me , including the Gamov region.
Signals produced by detectors were processed through standard electronics. The trigger signal was produced by the coincidence between S 1 (S 4) and the total R between S 4 5 6 (S 1 2 3). For the energy calibration we used a ^{12} beam on Au target to get carbon from elastic scattering, ^{12} beam on carbon target to get alphas for the ^{20}Ne α exit channel and a standard alpha source.

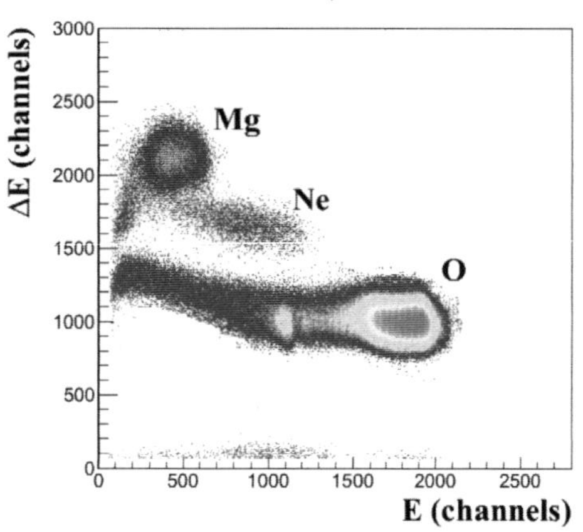

Figure 1. Δ matrix for neon identi cation.

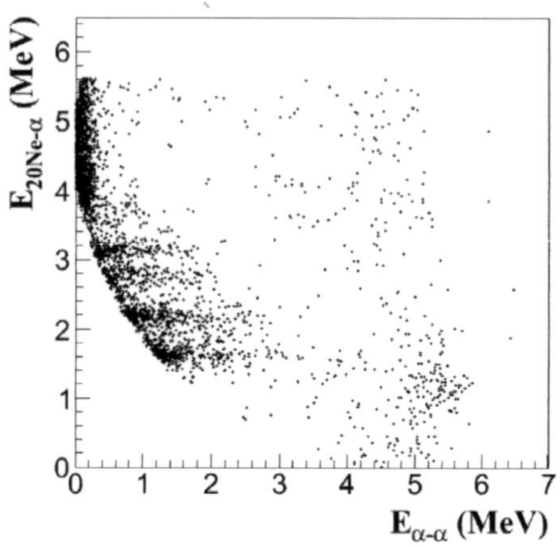

Figure 2. ^{20}Ne α relative energy $E_{^{20}Ne\alpha}$ versus α α relative energy $E_{\alpha\alpha}$

4. Data Analysis

To identify the reaction channel ^{16} (^{12} , α^{20}Ne)α of our interest, we performed the study of Δ matrices. In g.1 a typical experimental result is shown.

e can identify three loci related to oxygen (mainly due to the beam scattering) neon and magnesium. To select data of our interest a graphical cut was made around the neon locus. The experimental value for the three body process was reconstructed considering alpha particles detected on S 5 6 (S 2 3) and assuming that the third undetected particle is an alpha particle. The experimental values obtained 2.66 Me , 4.23 Me , corresponds with good approximation to the theoretical value of the ^{16} (^{12} , α^{20}Ne)α reaction with ^{20}Ne at g.s and 1st excited state respectively. This result con rms the correct selection of the reaction channel. For the following analysis we selected events related to ^{20}Ne in its g.s. In g.1 we can observe also that magnesium partially overlaps with the neon locus. This is partially due to the resolution of the detection system, on the other hand the loci are overlapped since we are in a energy range corresponding to the Bragg pic region. In particular, the contamination is evident for events corresponding to a inematic region with high probability for the uasi free process. For this reason we applied a procedure in order to remove spurious events avoiding to put sharp cuts. In order to study the reaction mechanism populating our three body channel, we reconstructed the relative energy matrices. In particular in g.2 we have ^{20}Ne alpha relative energy ($E_{^{20}Ne\alpha}$) versus alpha alpha relative energy ($E_{\alpha\alpha}$). The vertical locus for a x alpha alpha relative energy corresponds to the formation of the ^{8}Be g.s. It means that our exit channel is populated by a se uential process ^{16} ^{12} $\longrightarrow$ ^{20}Ne ^{8}Be* $\longrightarrow$ ^{20}Ne α α. The hori ontal loci for x $E_{^{20}Ne\alpha}$ energy correspond to the population of ^{24}Mg excited states.

ue to the l ^{12} α intercluster motion inside the ^{16} , the region where we expect to have the highest probability for the uasi free process corresponds to low values for momentum of the alpha particle spectator (p_s) (less than 1 Me c.) The study of the ^{20}Ne alpha relative energy as function of p_s indicates that the uasi free region experiences a very low statistics, moreover there is no evidence of ^{24}Mg excited states. This region corresponds to low energy ^{20}Ne area where a strong contamination of spurious events rules. So the low statistics could be due to wrong selection of the events in this region. n the other hand the ^{24}Mg excited states are populated in correspondence of high momenta. This result indicates that these levels are most probably populated by a se uential mechanism ^{16} ^{12} $\longrightarrow$ ^{24}Mg* α $\longrightarrow$ ^{20}Ne α α. In g.3 the spectrum of the ^{24}Mg exited states populated is shown. e can see ve levels.

Journal of Physics: Conference Series **703** (2016) 012025 doi:10.1088/1742-6596/703/1/012025

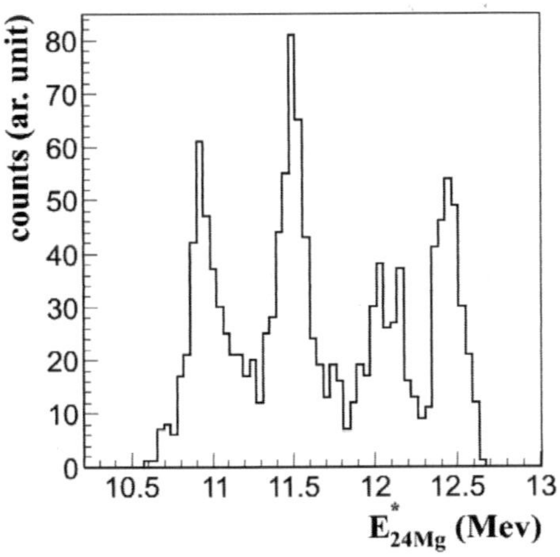

Figure 3. ^{24}Mg exited states populated in the present wor .

E^*_{24Mg} Me present wor	E^*_{24Mg} Me . Goldberg et *al.*	J^π
1 . 4		
11.5	11.52	2^+
12. 3	11. 6	2^+
12.16		
12.45	12.46	1^-

Table 1. ^{24}Mg exited states energy obtained in the present wor compared with the excited levels measured in . Goldberg *et al.* 15

ue to the low statistics, it was not possible to ma e a de nite J^π assignment. In tab.1 we report a comparison between our data and the results obtained by . Goldberg *et al.* 15 where the ^{24}Mg levels were observed via ^{20}Ne alpha resonant scattering.

5. Conclusion

In the framewor of the study of carbon carbon fusion reactions at astrophysical energy via T M, we have studied the ^{16} (^{12} , α^{20}Ne)α using for the rst time the ^{16} as Tro an orse nucleus. xperimental results obtained, in the experimental conditions described, have shown that three body exit channel is mainly populated by se uential decay mechanism while no clear evidence of uasi free process has been observed. The strong contamination of spurious events in the uasi free energy region could be the reason of wrong selection of uasi free events. ^{24}Mg excited states have been populated and compared with data present in literature. n the basis of this results a new experimental run has been performed with a di erent experimental set up in order to increase the resolution and avoid the contamination problems. Recently, we performed an experimental run where T M has been applied to the ^{14}N(^{12} , α ^{20}Ne)^2 three body reaction using for the rst time ^{14}N (^{12} $\oplus$ d cluster con guration) as Tro an orse nucleus. ata are currently being analy ed.

References

1 . . Rolfs and .S. Rodney, *Cauldrons in the Cosmos* (niversity of hicago ress, 1)
2 F. Strieder, J. hys.: onf. Ser. **202**, (2 1) 12 25.
3 T. Spillane et al., hys. Rev. ett. **98**, (2 7) 1225 1 4.
4 R. . ooper, A. . Steiner and .F. Brown, Astroph. J. **702**, (2) 66 671.
5 G. Baur et al., Nucl. hys. A **458**, (2 7) 1 .
6 . Spitaleri et al. hys. of Atomic Nuclei **74**, (2 11) 1725.
7 R. Tribble et al., Rep. rog. hys. **77**, (2 14) 1 6 1.
G. alvi et al. Nucl. hys. A **621** (1 7) 13 .
A. Tumino et al. Few Body Systems **54**, (2 13) 745.
1 . Spitaleri et al. hys. Rev. **90**, (2 14) 35 1.
11 A. Tumino et al. Astrophys. J. **785**, (2 14) 6.
12 M. . Sergi et al. hys. Rev. **91**, (2 15) 65 3.
13 S. herubini et al. hys. Rev. **92**, (2 15) 15 5.
14 M. Gulino et al. hys. Rev. **87**, (2 13) 12 1.
15 . Goldberg hys. Rev. **93**, (1 54) 7 .

Competition between the compound and the pre-compound emission processes in α-induced reactions at near astrophysical energy to well above it

Manoj Kumar Sharma1, Vijay Raj Sharma2, Abhiskek Yadav3, Pushpendra P Singh4, B. P. Singh2, and R. Prasad2

1 Department of Physics, Sri Varsheny College, Aligarh, India
2 Department of Physics, Aligarh Muslsim Uinversity, Aligarh, India
3 Interuniversity Accelerator Centre, New Delhi, India
4 Department of Physics, IIT, Ropar, India

E-mail: manojamu76@gmail.com

Abstract. The study of pre-compound emission in α-induced reactions, particularly at the low incident energies, is of considerable interest as the pre-compound emission is more likely to occur at higher energies. With a view to study the competition between the compound and the pre-compound emission processes in α-induced reactions at different energies and with different targets, a systematics for neutron emission channels in targets ^{51}V, ^{55}Mn, ^{93}Nb, ^{121,123}Sb and ^{141}Pr at energy ranging from astrophysical interest to well above it, has been developed.

The off-line γ-ray-spectrometry based activation technique has been adopted to measure the excitation functions. The experimental excitation functions have been analysed within the framework of the compound nucleus mechanism based on the Weisskopf-Ewing model and the pre-compound emission calculations based on the geometry dependent hybrid model. The analysis of the data shows that experimental excitation functions could be reproduced only when the pre-compound emission, simulated theoretically, is taken into account.

The strength of pre-compound emission process for each system has been obtained by deducing the pre-compound fraction. Analysis of data indicates that in α-induced reactions, the pre-compound emission process plays an important role, particularly at the low incident energies, where the pure compound nucleus process is likely to dominate.

1. Introduction

The mechanism of pre-equilibrium emission (PE), particularly at the lower incident energies, is of fundamental interest, since it is expected to occur at the higher excitation energies[1]. It reflects the dynamics of formation of an excited composite system and its evolution to the equilibrium states leading to formation of a compound nucleus (CN). A large number of nuclear reactions initiated by the light and heavy ions have been analysed in the frame work of the CN mechanism, where it is assumed that the excited composite system, formed by the absorption of incident ion lives long enough till thermodynamic equilibrium is established by the CN mechanism. However, both the experimental evidence and the intuition suggest that particle(s) emission may occur from the excited composite system formed by the absorption of the incident ion by the target nucleus before the establishment of thermodynamical equilibrium[2, 3, 4, 5]. These particles are called PE particles and the process as PE emission. Some of the important experimental signatures of the PE emission are; (i) the presence of a larger number of high-energy particles as compared to the spectrum predicted by the CN theory, (ii) forward-peaked angular distribution of the emitted particles, (iii) slowly decreasing tails of the excitation functions (EFs) etc.

The EFs data of low energy neutron capture are very useful in accurate determination of the astrophysical reaction rates for p and γ-processes[6, 7]. With the motivation to study the reaction mechanism in a consistent and systematic way, the measurements and analysis of EFs for (α,n) reactions on various targets having odd Z and odd A nuclei have been undertaken.

2. Measurements of Excitation functions

Spectroscopically pure targets of ^{55}Mn, ^{93}Nb and ^{141}Pr have been prepared by rolling method of desired thickness and are used as the samples. The experiments have been carried out at the Variable Energy Cyclotron Centre (VECC), Kolkata, India, using a collimated α-particle beam[1]. A stack consisting of target samples sandwiched between the aluminum degraders/catchers was irradiated for $\approx$12 h. The beam current $\approx$100 nA was monitored from the current integrator count rate. The calculations for average beam energy on a given target of the stack have been performed using the stopping power program SRIM[8]. The post-irradiation analysis has been carried out by using a high resolution large volume (100 c.c.) high-purity germanium (HPGe) detector coupled to an ORTEC PC based multichannel analyzer.

3. Analysis of excitation functions

In the present work, the code ALICE-91 developed by M. Blann[9], has been used to calculate the CN and the PE emission cross-sections. The CN calculations in this code are performed by using the Weisskopf-Ewing model[10], while, the PE component is simulated from the geometry dependent hybrid (GDH) model[11]. In this code, the level density parameter a,(a=A/K where, A be is the mass number of the composite nucleus and K is a free parameter), the initial exciton number n_0, and the mean free path multiplier $COST$ are some of the important parameters. The value of K=8, n_0=4 and $COST$=0.2 give satisfactorily reproduction of the experimental data. The experimentally measured EFs and calculated EFs with K=8, n_0=4 and $COST$=0.2 for the reactions ^{93}Nb(α, n)^{96}Tc and ^{141}Pr(α, n)^{144}Pm are shown in Fig. 1 (a-b), while for reaction ^{55}Mn(α, n)^{58}Co is shown in Fig. 2(a). In order to get systematics in PE emission (P_{FR}) on various targets, theoretical calculations of experimental data of other reactions ^{51}V(α, n)^{54}Mn, ^{121}Sb(α, n)^{124}I, and ^{123}Sb(α, n)^{126}I (which are not shown here) have also been performed by using code ALICE-91 with same set of parameters and are found in good agreement.

4. Pre-equilibrium fraction P_{FR}

The P_{FR} reflects the relative importance of the PE processes and is defined as the ratio of the difference of the cross-sections for (PCN+CN) emission and the CN emission [i.e., $\sigma_{(PCN+CN)} - \sigma_{(CN)}$] to the cross-section values of (PCN+CN) [i.e., $\sigma_{(PCN+CN)}$]. The deduced P_{FR} values are plotted as a function of the excitation energy for the reactions ^{51}V(α, n)^{54}Mn, ^{55}Mn(α, n)^{58}Co, ^{93}Nb(α, n)^{96}Tc, ^{121}Sb(α, n)^{124}I, ^{123}Sb(α, n)^{126}I, and ^{141}Pr(α,n)^{144}Pm and are shown in Fig. 2(b). The following conclusions are deduced from Fig. 2 (b); (i) P_{FR} for these reactions increases sharply with excitation energy for each target. The sharp increase in P_{FR} with the excitation energy shows that a very fast distribution of the excitation energy among the nucleons take a part in the PE emission, (ii) P_{FR} for presently studied reactions attains a maximum value, that particularly depends on the mass number of the target nucleus. The higher the mass of the target nuclei, lower the values of the excitation energies at which maxima occur, and (iii) there is an inconsistency in P_{FR} for similar target masses i.e., ^{51}V and ^{55}Mn. This inconsistency has been explained in terms of a shell effect.

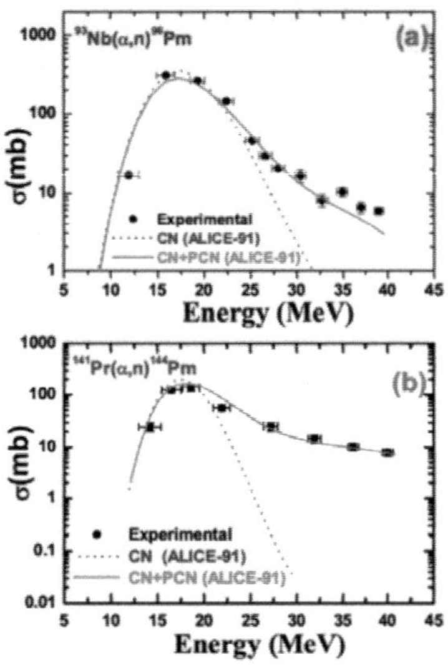

Figure 1. The experimentally measured and theoretically (ALICE-91) calculated EFs.

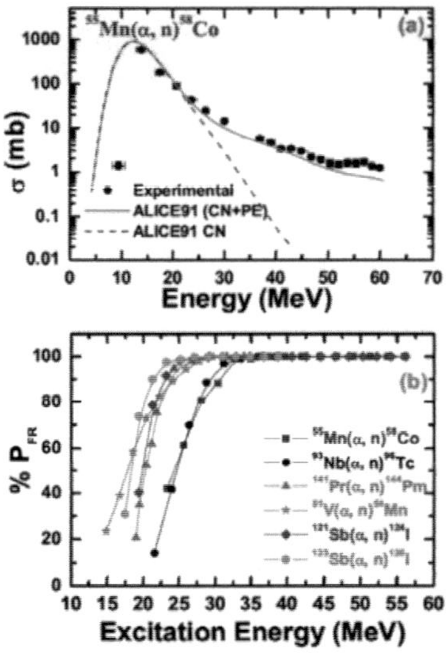

Figure 2. (a)The experimentally measured and theoretically (ALICE-91) calculated EFs, (b)Variation PE Fraction P_{FR} as a function of Excitation energy

5. Conclusions

The measurement and analysis EFs for reported reactions indicate that EFs could be reproduced only when the PE emission, simulated theoretically, is taken into account. The P_{FR} is found to depend on the excitation energy and structure of the target nuclei.

5.1. Acknowledgements

The author is thankful to the Director, VECC, Kolkata, India for extending all the facilities for carrying out the experiments and to Dr A. K. Dixit, Principal, S. V. College, Aligarh for his support.

6. References

[1] Singh B P, Bhardwaj H D, and Prasad R 1991 *Can. J. of Phys.* **69** 1376
[2] Avrigeanu M, Avrigeanu V, Bem P, Fischer U, Honusek M, Katovsky K, . Manailescu C, Mrazek J, Simeckova E, and Zavorka L 2014 it Phys. Rev. C **89** 044613
[3] Kim K, Kim G N, Naik H, Zaman M, Yang S -C, Song T-Y, Guin R and Das S K 2015 *Nucl. Phy. A.* **935** 65
[4] Sharma M K,Bhardwaj H D, Unnati, Singh P P, Singh B P, and Prasad R 2007 *Ero Phys. J.* **31** 43
[5] Singh B P, Sharma M K, Muthafa M M, Bhardwaj H D and Prasad R 2006 *Nucl. Inst. and Methods in Phys. Res. A.* **562** 717
[6] Sauerwein A, Becker H -W, Dombrowski H, Elvers M, Endres J, Giesen U, Hasper J, Hennig A, Netterdon L, Rauscher T, Rogalla D, Zell K O and Zilges A 2011 *Phys. Rev. C.* **C84** 045808
[7] Gyurky Gy, Vakulenko M, Fulop Zs, Halasz Z, Kiss G G, Somorjai E and Szucs T 2014 *Nucl. Phy. A* **992** 112
[8] http://www.srim.org
[9] Blann M 1991 *NEA Data Bank, Gif-sur-Yvette, France.* **Report PSR-146**
[10] Weisskopf V F and Ewing D H 1940 *Phys. Rev.* **57** 472
[11] Blann M 1972 *Phys. Rev. Lett.* **28** 757

Journal of Physics: Conference Series **703** (2016) 012027 doi:10.1088/1742-6596/703/1/012027

New measurement of the ^{10}B(n,α)^{7}Li through the Trojan Horse Method

Roberta Spartá
Laboratori Nazionali del Sud, INFN, Catania, Italy
E-mail: `rsparta@lns.infn.it`

Abstract. ^{10}B(n,α)^{7}Li reaction cross section has been measured using the Trojan Horse method, with the specific aim to separate the α_1 contribution (coming from the first ^{7}Li excited level) by the α_0 (related to the ^{7}Li ground state), using a very thin target. Preliminary results are shown of the three-body ^{10}B(d,α^{7}Li)H cross section.

1. Introduction

Studying the ^{10}B(n,α)^{7}Li cross section has become always more important in these last years, because of the importance of its knowledge for applied physics (it is considered to be a standard in neutron induced cross section [1]). Applications regard nuclear energy production in actual and future plants (it is widely used to monitor the neutron flux through ionization chambers filled with ^{10}B) [2] and medical cure. Indeed, it is used for the cure of melanomas and rheumatoid arthritis, with a treatment called Boron Neutron Capture Therapy (BNCT) [3] [4]; moreover, this reaction is helpful to study negative effects of the α particles doses on the human body [5].

This reaction has already been measured using Trojan Horse Method [6], but now a new measurement has been performed in order to separate the the two main contribution to this cross section, coming from the ground state and from the first excited state (energy 0.477 MeV) of the ^{7}Li.

2. Experimental apparatus

The experimental run has been performed in 2014 in Laboratori Nazionali del Sud (LNS, Catania), using a 28 MeV ^{10}B beam impinging on a CD_2 target, to measure the ^{10}B(d,α ^{7}Li)H cross section. Outcoming particles of interest (α and ^{7}Li) have been detected by four PSD detectors, symmetrically placed inside the chamber, shown in fig. 1. Details are reported in [7]. To get the two α contributions separated, it has been used a very this target (56 μg/cm^2), because it helped in minimizing the energy straggling and loss.

3. Three-body reaction selection

Once α and ^{7}Li produced have been discerned via their energy loss in isobutane inside the ionization chambers, the Q-value spectrum, reported in [7], clearly shows two separated peaks related to the two channels desired. From now on, it will be referred to α_0 and α_1 data meaning data under the two peaks in these Q-value spectra. The corresponding kinematical loci are in very nice agreement with the Monte Carlo simulations (fig. 2).

Figure 1. The Camera2000 in LNS arranged to measure ^{10}B(d,α ^{7}Li)H with PSD and Ionization Chambers.

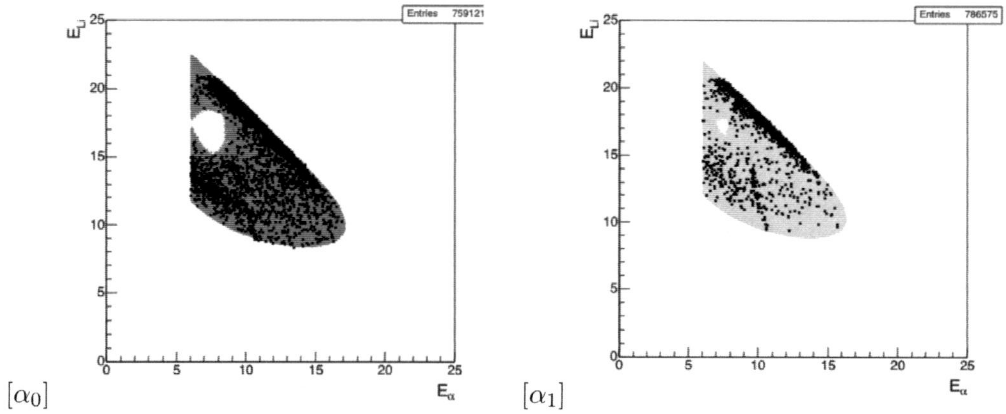

Figure 2. Kinematical loci (E_{7Li} vs. E_α, black points) are in agreement with the Monte Carlo simulations, respectively in green and red for α_1 and α_0 data.

4. Quasi-free mechanism selection

As common procedure for Trojan Horse Experiment, the spectator (a proton, in this case) momentum distribution must be considered to select the quasi-free mechanism. In fig. 3 it is evident that data, for both the channels, show a Hultn function shape, that is the momentum distribution function of the proton inside the deuteron for the l=0 wave, peaked at $p_s = 0$ MeV/c. This intends that proton acts as a spectator to the two body process, selecting data under this function, whose full width half maximum is in agreement with what expected experimentally, considering the transferred momentum [8]. This confirms the presence of the quasi-free mechanism, leading the analysis to the following step, the cross section extraction.

5. Preliminary results

The preliminary three body cross section, considering only data with $p_s < |40|$ MeV/c, is shown in fig. 4, for both channels. In conclusion, the separation of the two α contributions (the primary aim of this measurement) has been reached, propelling the further analysis to the final result of the two body reaction cross section.

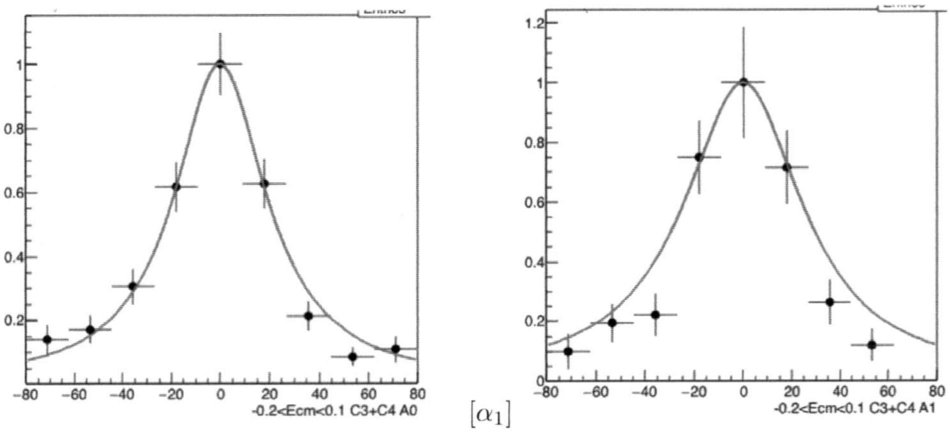

$[\alpha_0]$ $[\alpha_1]$

Figure 3. Spectator momentum distribution (black points) following the Hultn function, indicating the quasi-free mechanism presence.

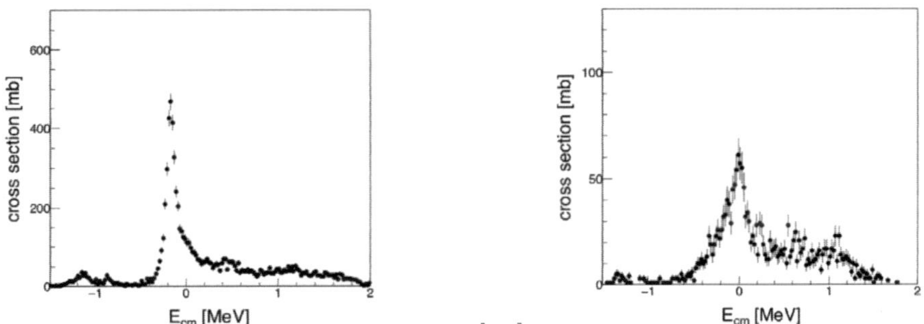

$[\alpha_0]$ $[\alpha_1]$

Figure 4. Three body reaction cross section for α_0 and α_1 channel.

5.1. Acknowledgments
The authors acknowledge the support of the Italian Ministry of Education, University and Research under grant Progetto Premiale ASTROF.MIUR PFE, CD 19/12/2012 n. 12609.

6. References

[1] https://www-nds.iaea.org/standards/ International Atomic Energy Agency - IAEA
[2] C. Rangacharyulu, *Physics of Nuclear Radiations: Concepts, Techniques and Applications*, CRC Press (2014)
[3] R.F. Barth et al., *Boron neutron capture therapy for cancer*, Cancer, 70-12 (1992)
[4] J. C. Yanch et al., *Boron neutron capture synovectomy: Treatment of rheumatoid arthritis based on* $^{10}B(n,\alpha)^7Li$ *nuclear reaction*, Med. Phys. 26, 364 (1999)
[5] N. Vulpis, *Chromosome aberrations induced in human peripheral blood lymphocytes using heavy particles from* $^{10}B(n,\alpha)^7Li$ *reaction*, Mutation Research/Fundamental and Molecular Mechanism of Mutagenesis, 18-1 (1973)
[6] L. Lamia et al., *Indirect study of (p,α) and (n,α) reactions induced on boron isotopes*, Nuovo Cimento, 31 C (2008)
[7] R. Spartá et al., *Indirect measurement of the* $^{10}B(n,\alpha)^7Li$ *to discern the* α_0 *e* α_1 *contribution to the total cross section*, LNS Activity Report (2009)
[8] R. G. Pizzone et al., *Effects of distortion of the intercluster motion in 2H, 3He, 3H, 6Li, and 9Be on Trojan horse applications*, Phys. Rev. C, 80-025807 (2009)

Effects of the s-process on Fe-group elements in meteorites

O Trippella[1,2], M Busso[1,2], G J Wasserburg[3], S Palmerini[1,2], F Frondini[1,2], M Petrelli[1,2], and A Zucchini[1,2]

[1] Department of Physics and Geology, University of Perugia, Perugia, Italy
[2] Istituto Nazionale di Fisica Nucleare, Section of Perugia, Perugia, Italy
[3] Lunatic Asylum, Geological and Planetary Sciences, California Institute of Technology, Pasadena, USA

E-mail: oscar.trippella@fisica.unipg.it

Abstract. In the present paper we investigate the possible connection between *s*-process nucleosynthesis occurring during the asymptotic giant branch (AGB) phase of low-mass stars (LMS) and the isotopic anomalies of the "Fe-group" elements observed in several macroscopic samples of meteorites or in grains formed as circumstellar condensates (hereafter CIRCONs). The available measurements of chromium, iron, and nickel are well reproduced by stellar models, which account for the largest shifts in the heaviest isotopes of each element: in particular ^{54}Cr, ^{58}Fe, and ^{64}Ni. Moreover, many circumstellar condensates reflect ^{50}Ti excesses and some production of ^{46,47,49}Ti, as predicted by slow-neutron captures in AGB stars. Nevertheless, some difficulties are found in comparing theoretical calculations of *s*-process nucleosynthesis with calcium, silicon, and zinc isotopic anomalies.

1. Introduction

The production of elements generally considered to be members of the iron-group (such as Ti, Cr, Fe, Ni, and Zn) is commonly attributed to supernovae (SNe), both core collapse and SNeIa. The isotopes of these elements can now be traced in pristine meteorites with high accuracy, as in the last decades several analysis techniques have been developed allowing to measure anomalies for the abundances of elements and isotopes at the level of a few parts in ten thousand (ϵ units). These isotopic effects are common in calcium-aluminum-rich inclusions (CAIs) in meteorites, but are also wide spread at lower levels in "bulk" samples of different groups of meteorites. In this context, we suggest that slow neutron captures, the so-called *s*-process occurring in the asymptotic giant branch phases (AGB) low mass stars (LMS), modify the original abundance of these elements collected from the interstellar medium by the star, by adding more neutrons, thus altering the isotopic admixture in favour of heavier isotopes [1].

In order to obtain a direct comparison with the meteoritic measurements, the enrichment factors $E^{\odot}(i,k)$ for isotopic abundances of the "Fe-group" elements are calculated as follow:

$$E^{\odot}(i,k) = \left[\left(\frac{N(i)}{N(k)}\right)^{\star} / \left(\frac{N(i)}{N(k)}\right)^{\odot} \right] - 1, \quad (1)$$

where an initial solar abundance [2] was assumed and "k" is a reference nucleus which is of high abundance. The composition of the envelopes of 1.5 and 3 $M_{\odot}$ model stars, for solar and

1/3 solar metallicity, were computed till the end of the AGB phase, using our post-process nucleosynehsis code NEWTON and adopting an extended ^{13}C-pocket as described in [3].

2. Cr, Ni, and Fe

The enrichment factors in isotopic abundances for Cr, Fe, and Ni are shown in panels A, B, and C, respectively of Figure 1. The *s*-process nucleosynthesis calculations suggest a large positive shift for ^{54}Cr, ^{58}Fe, and ^{64}Ni (representing the heaviest stable isotope of each element), while only limited variations are expected for other nuclei and in some cases, as for ^{50}Cr and ^{54}Fe, even a negative shift is predicted.

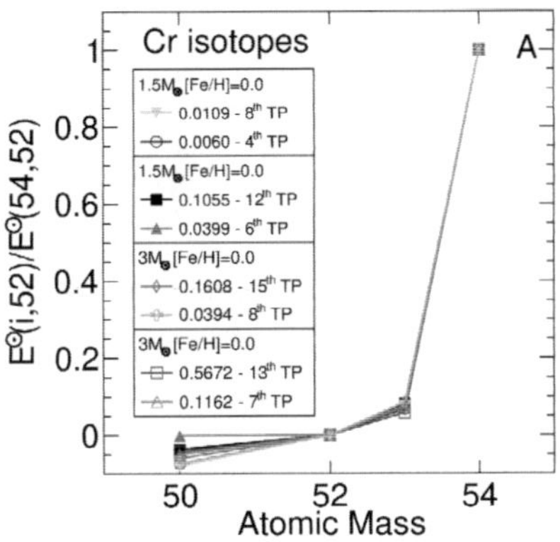

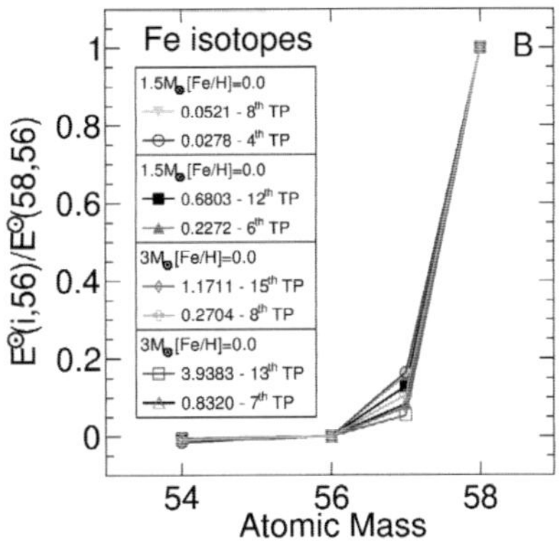

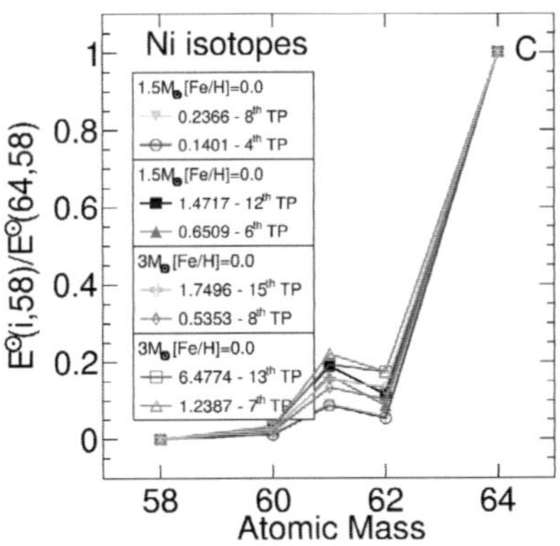

Figure 1. Cr, Fe, and Ni enrichment factors for the envelopes of model AGB stars with mass of 1.5 and 3 $M_{\odot}$, having a solar or 1/3 solar initial composition. Results for an intermediate and the final thermal instability of each stellar model are shown with different colors for both marker and line. In the inside legend we also indicate the absolute values of $E^{\odot}(i,k)$.

Starting from the comparison of results in Figure 1 with data measured from those samples which are associated with AGB stars (e.g. [4] in the case of Ni), our suggestion is that the "iron group" elements mentioned above were originally made in SNe, but their products were subsequently reprocessed in AGB stars, resulting in the observed effects. Moreover, it turns out that the magnitude of the enrichment factors is dependent on the size of the ^{13}C-pocket, the stellar mass, the initial composition, and the stage of evolution of the star.

3. Titanium isotopes

In Figure 2 we collect a wide selection of Ti isotopic data on main stream CIRCONS compared to AGB *s*-process patterns. We observe that slow neutron captures calculations show a typical "J" pattern [1, 5] because of smaller $E^{\odot}(46,48)$ shifts with respect to $E^{\odot}(50,48)$ ones.

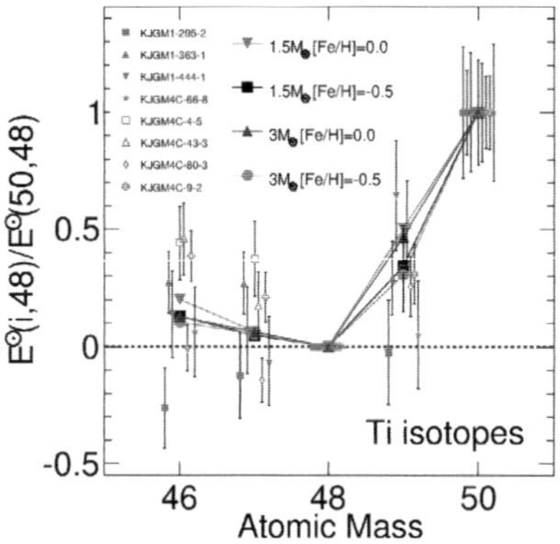

Figure 2. Comparison between Ti isotopic data (grey points) on main stream CIRCONS ([6], [7], and [8]) and s-process predictions in AGB-LMS. Markers and lines of different colors refer to calculations regarding stars of 1.5 and 3 $M_\odot$, solar and 1/3 solar metallicity. For the sake of comparison, measurement data and s-process theoretical predictions are normalized to $\delta(50,48)$ and $E^\odot(50,48)$, respectively.

There are abundant isotopic data (see grey dots in Figure 2 and references therein) on titanium and many circumstellar condensates reflect ^{50}Ti excesses and some production of ^{46,47,49}Ti, as suggested by AGB neutron captures. From these observations we conclude that a large fraction of these grains show clear evidence of s-processing. However, there are also cases showing a disagreement with higher shifts in lighter nuclei, giving the image of a "V" pattern.

4. Conclusions

We computed the effects of slow neutron capture nucleosynthesis on Cr, Fe, and Ni through 1.5 $M_\odot$ and 3 $M_\odot$ AGB models, adopting solar and 1/3 solar metallicities. All cases produce excesses of ^{54}Cr, ^{58}Fe, and ^{64}Ni, while the other isotopes are little altered; hence, the observations made in recent years in meteorites may be explained by AGB processing.

The case of titanium is roughly similar and it is possible to correlate the s-processing predictions with data from those circumstellar condensates that are usually associated with AGB sources (those belonging to the so-called "main stream").

On the other hand, concerning the effects of s-processing on silicon, calcium, and zinc, the scenario is more complex. For Si, the main effects are due to variations in the local ISM from different SNe sources. Calcium still represents an open problem because calculations suggest large anomalies on ^{46}Ca, but no observations support this. In general, the Ca shifts in s-processing are relatively weak as compared to Ti, Cr, Fe, and Ni. At the end, no clear explanation are possible for Zn because only limited measurements are available and they don't show large shifts for ^{68}Zn as predicted by nuclesosynthesis calculations.

5. Acknowledgments

This research was supported by the Department of Physics and Geology of the University of Perugia (through the grant named "From Rocks to Stars") and the INFN, Section of Perugia through the funds of its Group III.

References

[1] Wasserburg G J, Trippella O and Busso M 2015 *ApJ* **805** 7
[2] Lodders K, Palme H and Gail H P 2009 *Landolt Börnstein* 44
[3] Trippella O, Busso M, Maiorca E, Käppeler F and Palmerini S 2014 *ApJ* **787** 41
[4] Chen J H and Papanastassiou D A 2014 *LPSC* (*Lunar and Planetary Inst. Technical Report* vol 45) p 2327
[5] Choi B G, Huss G R, Wasserburg G J and Gallino R 1998 *Science* **282** 1284
[6] Amari S, Nittler L R, Zinner E, Lodders K and Lewis R S 2001 *ApJ* **559** 463–483
[7] Amari S, Nittler L R, Zinner E, Gallino R, Lugaro M and Lewis R S 2001 *ApJ* **546** 248–266
[8] Huss G R and Smith J B 2007 *Meteoritics and Planetary Science* **42** 1055–1075

Experimental study of the $^{13}C+^{12}C$ fusion reaction at deep sub-barrier energies

D. Tudor[1], A.I. Chilug[1], M. Straticiuc[1], L. Trache[1], D. Chesneanu[1], S. Toma[1], D. G. Ghita[1], I. Burducea[1], R. Margineanu[1], A. Pantelica[1], C. Gomoiu[1], N.T. Zhang[2], X. Tang[3], Y.J. Li[3]

[1] *IFIN-HH, Bucharest-Magurele, Romania*
[2] *N.T. Zhang, X. Tang - Institute of Modern Physics, Lanzhou, China*
[3] *Y.J. Li - CIAE, Beijing, China*

Abstract. Heavy-ion fusion reactions between light nuclei such as carbon and oxygen isotopes have been studied because of their significance for a wide variety of stellar burning scenarios. One important stellar reaction is $^{12}C+^{12}C$, but it is difficult to measure it in the Gamow window because of very low cross sections and several resonances occurring. Hints can be obtained from the study of $^{13}C+^{12}C$ reaction. We have measured this process by an activation method for energies down to E_{cm}=2.5 MeV using ^{13}C beams from the Bucharest 3 MV tandetron and gamma-ray deactivation measurements in our low and ultralow background laboratories, the latter located in a salt mine about 100 km north of Bucharest. Results obtained so far are shown and discussed in connection with the possibility to go even further down in energy and with the interpretation of the reaction mechanism at such deep sub-barrier energies.

1. Introduction

One of the important questions in nuclear astrophysics is the carbon burning scenario. This process represents the third stage of stellar evolution of massive stars. Until now fusion reaction have only been measured at energies well above the region of astrophysical interest because of the extremely low cross section and signal/background ratio. In stellar environments the reaction rates are estimated by extrapolating measurements done at higher energies, extrapolations that imply a certain degree of uncertainty. For the $^{12}C+^{12}C$ fusion reaction the situation is more complicated because of the resonances occuring below the Coulomb barrier. Therefore direct measurements at the Gamow window energies are essential, but very difficult to carry out due to the background from the cosmic rays, terrestrial environment etc.

We have established a program at IFIN-HH Bucharest-Magurele to test the possibility to make direct measurements for nuclear astrophysics using the new 3 MV tandetron accelerator [1] and an ultra-low background laboratory [2] situated in a salt mine at about 2.5 hours drive north of Bucharest. Therefore improvements can be made using irradiation de-activation sequences: we irradiate probes at the new 3 MV tandetron accelerator and move the probes for de-activation measurements in the ultra-low background laboratory in the salt mine. The preliminary results are presented here. The measurement we report in this paper has been performed in collaboration with the IMP Lanzhou.

2. ^{13}C+^{12}C Experiment

The resonances from the ^{12}C+^{12}C fusion reaction make it very difficult to measure, so to be able to test the predictive power of various models and establish a reliable upper limit for the cross section, we studied the ^{13}C+^{12}C fusion cross section at deep sub-barrier energies. After initial tests of accelerator performances, we have proven that it has high and stable beam current in the range of tens of μA for negative ions (^{12}C, ^{13}C, ^{28}Si, ^{197}Au). The machine has a maximum voltage of 3.2 MV, and it can run as low as 200 kV. We concluded that we could be competitive to perform the study of reactions induced by alphas and light ions (0.2-1 MeV per nucleon).

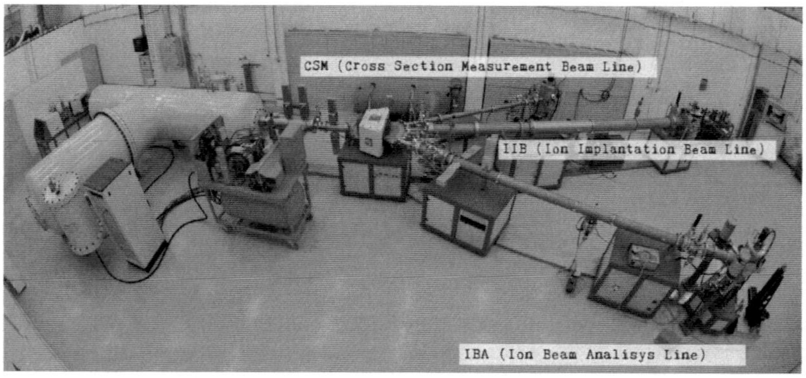

Figure 1. 3 MV Tandetron Accelerator at IFIN-HH [4].

The reaction we chose ^{13}C+^{12}C has the advantage that leads to an activation product with a half-life allowing for the transportation and efficient de-activation measurement. Therefore, our interest was focused on the proton evaporation channel ^{12}C(^{13}C,p)^{24}Na. The experiment had been performed in October 2014. The ^{13}C beam at laboratory energy range of $E_{lab} = 6.8 - 5.2$ MeV ($E_{cm} = 3.2 - 2.5$ MeV), with steps of 0.2 MeV, impinged on 1 mm thick natural carbon targets. Intensities in the range of 2-8 pμA were used in different runs. The irradiation chamber was electrically isolated, acting as a Faraday cup for current integration.

Thick target yield for the ^{12}C(^{13}C,p)^{24}Na fusion reaction was determined through the measurement of the gamma-ray yield following the beta-decay of ^{24}Na ($T_{1/2}$=15 h) at the GammaSpec laboratory at the ground level (in IFIN-HH) and the ultra-low backround laboratory μBq (in the salt mine at Slanic). At the μBq the background decreased almost 4000 times (Fig. 2). In these laboratories the cascading gamma rays (1369 and 2754 keV) were detected with HPGe detectors with 30% relative efficiency (at GammaSpec) and 120% (at μBq in the salt mine) [2]. Their calibration is well known.

The ^{12}C(^{13}C,p)^{24}Na cross sections was calculated from the extracted thick target yield Y(E), using the following equation:

$$\sigma(E) = \frac{1}{N_\nu}\frac{dY(E)}{dE}\frac{dE}{dx}, \tag{1}$$

where N_ν is the number density of target nuclei present in the target and dE/dx is the stopping power.

As you can see (Fig. 3) we went down to the lowest energy ever reached E_{cm}=2.5 MeV and have obtained a good agreement with the past experiments [3-5]. Proton-evaporation cross section were determined and statistical model calculations were used to evaluate the total reaction cross section.

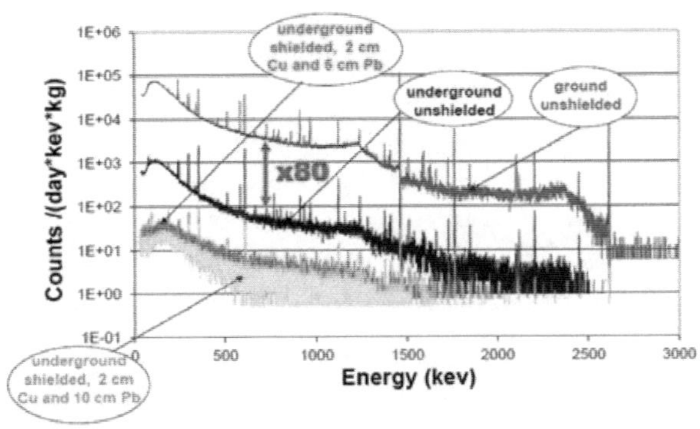

Figure 2. Background from the salt mine (μBq laboratory)[2].

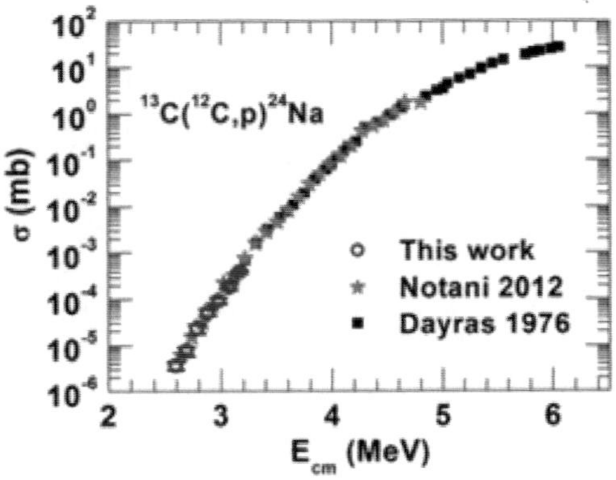

Figure 3. Preliminary cross section of ^{12}C(^{13}C,p)^{24}Na from our experiment and results from the previous experiments.

3. Summary

We have made a number of irradiations at the 3 MV tandetron and measured the ^{12}C(^{13}C,p)^{24}Na thick-target yield. The activation measurements were done at the low background laboratory GammaSpec and the ultra-low background laboratory μBq in the Slanic-Prahova salt mine. We went down to the lowest energy ever reached for this reaction (E_{cm}=2.5 MeV), where the cross section was of 4 nb. We have obtained a good agreement with the past experiments and for this year experiment (October 2015) we plan to go to even lower energies. For the ground level measurements (at IFIN-HH) we plan to perform beta-gamma coincidences to further clean the background.

Acknowledgments

Part of this work was supported by the Romanian National Authority for Scientific Research under the project Nuclear astrophysics studies at IFIN-HH, PNII-ID-PCE-2012-4-0299.

References

[1] I. Burducea, M. Straticiuc, D.G. Ghita, D.V. Mosu, C.I. Calinescu, N.C. Podaru, D.J.W. Mous, I. Ursu, N.V. Zamfir,Nuclear Instruments and Methods in Physics Research B 359, 1219, 2015.

[2] R. Margineanu, C. Simion, S. Bercea, O.G. Duliu, D. Gheorghiu, A. Stochioiu, M. Matei, Applied Radiation and Isotopes 66, 1501-1506, 2008.

[3] R.A.Dayras, R.G. Stokstad, Z.E.Switkowski and R.M.Wieland, Nuclear Physics A 625, 152-188, 1976.

[4] X. Fang et al., Journal of Physics: Conference Series 420, 012151, 2013.

[5] M. Notani, X.D. Tang et al., Physcal Review C 85, 014607, 2012.